CHEMICAL
REFERENCE MATERIALS

SETTING THE STANDARDS FOR OCEAN SCIENCE

Committee on Reference Materials for Ocean Science

Ocean Studies Board
Division on Earth and Life Studies

NATIONAL RESEARCH COUNCIL
OF THE NATIONAL ACADEMIES

THE NATIONAL ACADEMIES PRESS
Washington, D.C.
www.nap.edu

THE NATIONAL ACADEMIES PRESS • 500 Fifth Street, NW • Washington, DC 20001

NOTICE: The project that is the subject of this report was approved by the Governing Board of the National Research Council, whose members are drawn from the councils of the National Academy of Sciences, the National Academy of Engineering, and the Institute of Medicine. The members of the committee responsible for the report were chosen for their special competences and with regard for appropriate balance.

This material is based upon work supported by the National Science Foundation (NSF grant number OCE-0096792). Any opinions, findings, and conclusions or recommendations expressed in this material are those of the authors and do not necessarily reflect the views of the National Science Foundation.

International Standard Book Number: 0-309-08500-4

Additional copies of this report are available from:

The National Academies Press
500 Fifth Street, N.W.
Box 285
Washington, DC 20055
(800) 624-6242
(202) 334-3313 (in the Washington metropolitan area)
http://www.nap.edu

Copyright 2002 by the National Academy of Sciences. All rights reserved.

Printed in the United States of America.

THE NATIONAL ACADEMIES
Advisers to the Nation on Science, Engineering, and Medicine

The **National Academy of Sciences** is a private, nonprofit, self-perpetuating society of distinguished scholars engaged in scientific and engineering research, dedicated to the furtherance of science and technology and to their use for the general welfare. Upon the authority of the charter granted to it by the Congress in 1863, the Academy has a mandate that requires it to advise the federal government on scientific and technical matters. Dr. Bruce M. Alberts is president of the National Academy of Sciences.

The **National Academy of Engineering** was established in 1964, under the charter of the National Academy of Sciences, as a parallel organization of outstanding engineers. It is autonomous in its administration and in the selection of its members, sharing with the National Academy of Sciences the responsibility for advising the federal government. The National Academy of Engineering also sponsors engineering programs aimed at meeting national needs, encourages education and research, and recognizes the superior achievements of engineers. Dr. Wm. A. Wulf is president of the National Academy of Engineering.

The **Institute of Medicine** was established in 1970 by the National Academy of Sciences to secure the services of eminent members of appropriate professions in the examination of policy matters pertaining to the health of the public. The Institute acts under the responsibility given to the National Academy of Sciences by its congressional charter to be an adviser to the federal government and, upon its own initiative, to identify issues of medical care, research, and education. Dr. Harvey V. Fineberg is president of the Institute of Medicine.

The **National Research Council** was organized by the National Academy of Sciences in 1916 to associate the broad community of science and technology with the Academy's purposes of furthering knowledge and advising the federal government. Functioning in accordance with general policies determined by the Academy, the Council has become the principal operating agency of both the National Academy of Sciences and the National Academy of Engineering in providing services to the government, the public, and the scientific and engineering communities. The Council is administered jointly by both Academies and the Institute of Medicine. Dr. Bruce M. Alberts and Dr. Wm. A. Wulf are chair and vice chair, respectively, of the National Research Council.

www.national-academies.org

This report is dedicated to the memory of our colleague
John Hedges
and his many contributions
to oceanography and organic geochemistry.
(1946-2002)

---

* Deceased July 2002

SHIREL SMITH, Office Manager
JODI BACHIM, Senior Project Assistant
NANCY CAPUTO, Senior Project Assistant
DENISE GREENE, Senior Project Assistant
DARLA KOENIG, Senior Project Assistant
JULIE PULLEY, Project Assistant
ALISON SCHRUM, Project Assistant

Acknowledgments

The committee would like to acknowledge the contributions and support of its sponsor, the National Science Foundation. This report was also greatly enhanced by the input of the invited representatives from government agencies with experience in oceanic reference materials who gave talks at the planning meetings: Don Rice, National Science Foundation; Adriana Cantillo, National Oceanic and Atmospheric Administration; John Fassett, National Institute of Standards and Technology; and Scott Willy, National Research Council of Canada. Input was also solicited through e-mail from a broad cross-section of the marine community world-wide, with help from the American Society of Limnology and Oceanography.

This report has been reviewed in draft form by individuals chosen for their diverse perspectives and technical expertise, in accordance with procedures approved by the National Research Council's Report Review Committee. The purpose of this independent review is to provide candid and critical comments that will assist the institution in making its published report as sound as possible and to ensure that the report meets institutional standards for objectivity, evidence, and responsiveness to the study charge. The review comments and draft manuscript remain confidential to protect the integrity of the deliberative process. We wish to thank the following individuals for their review of this report: Dr. Richard T. Barber (Duke University), Dr. Edward Boyle (Massachusetts Institute of Technology), Dr. Thomas S. Bianchi (Tulane University), Dr. Katherine H. Freeman (Penn State), Dr. Dennis A. Hansell (University of Miami), Dr. Susan Libes (Coastal Carolina University), Dr. Steven E. Lohrenz (University of Southern Mississippi), Dr. Jay Pinckney (Texas A&M University), and Dr. Thomas Torgersen (University of Connecticut).

Although the reviewers listed above have provided many constructive comments and suggestions, they were not asked to endorse the conclusions or recommendations nor did they see the final draft of the report before its release. The review of this report was overseen by Dr. Kenneth H. Brink, Woods Hole Oceanographic Institution. Appointed by the National Research Council, he was responsible for making certain that an independent examination of this report was carried out in accordance with institutional procedures and that all review comments were carefully considered. Responsibility for the final content of this report rests entirely with the authoring committee and the institution. Maggie Sheer provided valuable assistance with copy-editing. The artwork and cover were designed by Van Nguyen.

Preface

Chemical Reference Materials: Setting the Standards for Ocean Science is part of an evolving body of work being conducted by scientists and research sponsors interested in ensuring the quality control of oceanographic data. Chemical data collected during ongoing and future global oceanographic studies and time-series efforts must be comparable over time and among laboratories. A wide range of scientific opportunities will result from such long-term observations, such as a better understanding of the role of ocean chemistry in climate dynamics; also improved stewardship of the ocean's natural resources. The large investment of time, money, and equipment needed for such chemical oceanographic measurements demands that the data collected be of the highest quality achievable. Chemical reference materials play a critical role in the verification of the quality of these measurements. To this end, the National Research Council Committee on Reference Materials for Ocean Science (Appendix A) was charged with the difficult tasks of identifying the most critically needed reference materials, and recommending the most appropriate approaches for their development. The committee gave careful consideration to keeping their recommendations within the context of current and future oceanographic efforts throughout this process.

Committee members were chosen for their wide variety of scientific expertise and experience in production and certification of reference materials. In addition, members with proficiency in the use of reference materials for the analysis of trace metals, radioisotopes, nutrients, carbon, and organic matter were represented. The committee met on four separate occasions to discuss and to plan this report. One of these meetings was a workshop held in September of 2001 in Islamorada, Florida at which about 30 invitees from the ocean science community (Appendix B) listened to keynote presentations, and discussed which reference materials, if available, would enhance the ability of ocean scientists to address key research topics. In addition, workshop participants were asked to identify which materials they felt represented the highest priority for development and research. Workshop participants, posters, and discussions helped set the stage for the fruitful committee discussions that followed. The committee also relied on written comments provided by workshop participants, on an email survey of members of the American Society of Limnology and Oceanography, and on the National Science Foundation's report on the *Future of Ocean Chemistry in the U.S.* (1999), which set research priorities in marine chemistry.

As this report went to press, the committee was saddened by the unexpected death of a committee member, Dr. John Hedges. Dr. Hedges' extensive and thoughtful input to this report reflected his deep interest in this topic and his hopes that this report would meaningfully further the use of reference materials in the ocean sciences. His death is a great loss to the many individuals who knew him personally and to the chemical oceanography community as a whole.

Committee on Reference
Materials for Ocean Science

Contents

EXECUTIVE SUMMARY   1

1 INTRODUCTION   7
   Background to the Study, 7
   Benefits of Chemical Reference Materials to Ocean Science, 11
   Report Structure, 14

2 IMPROVING CHEMICAL OCEANOGRAPHIC DATA   17
   Matrix Dependence of Reference Materials, 17
   How Reference Materials Work, 18
   Limitations of Reference Materials, 20
   Benefits of Reference Materials, 22

3 SEAWATER   29
   Nutrients, 29
   Trace Metals, 33
   Radionuclides, 37
   Carbon Isotopes in Dissolved Inorganic Carbon (DIC), 41
   Dissolved Organic Matter (DOM), 43
   Dissolved Gases, 46

4 CHEMICAL REFERENCE MATERIALS FOR THE
ANALYSIS OF PARTICULATE AND SEDIMENT SAMPLES 47
Rationale for Sediment and Particulate Matter Analysis, 47
Influence of Matrix Composition on Chemical Determinations, 57
Reference Materials Currently Available for the Analysis of
   Sediment and Particulate Samples, 66
Recommended Reference Materials, 72
Potential Long-Term Needs for Additional Reference Materials, 75

5 PRODUCTION AND DISTRIBUTION OF CHEMICAL
REFERENCE MATERIALS 77
Introduction, 77
Requirements of Reference Materials, 78
Reference Material and Certified Reference Material
   Production, 78
Methods Employed to Characterize Reference Materials and
   Certified Reference Materials, 81
Preparation of Recommended New Reference Materials and
   Certified Reference Materials for Ocean Science, 82
Costs of Producing and Distributing Reference Materials and
   Certified Reference Materials, 85
A Strategy for the Production of New Reference Materials for
   the Ocean Sciences, 86
Education, 87

6 CONCLUSIONS AND RECOMMENDATIONS 89
Recommendations for Reference Materials for Ocean Science, 90
Recommendations for Community Participation, 95
Statement of Top Priorities, 96

REFERENCES CITED 99

APPENDIXES
A COMMITTEE AND STAFF BIOGRAPHIES 111
B WORKSHOP PARTICIPANTS 115
C GLOSSARY 116
D ACRONYM LIST AND CHEMICAL TERMINOLOGY 122
E REFERENCE MATERIALS LISTED WITHIN THIS REPORT 126

Executive Summary

The accuracy of chemical oceanographic measurements depends on calibration against reference materials to ensure comparability over time and among laboratories. Several key parameters lack reference materials for measurements in seawater, particles in the water column, and sediments. Without reference materials it is difficult to produce the reliable data sets or long-term baseline studies that are essential to verify global change and oceanic stability. This report identifies the most urgently required chemical reference materials based on key themes for oceanographic research and provides suggestions as to how they can be developed within realistic cost constraints.

Chemical analyses of seawater are uniquely difficult given the poorly known speciation and the low concentration of many of the analytes of interest. Analyses of suspended and sedimentary marine particulate materials present their own distinct challenges, primarily due to potential interference by predominant mineral phases of different types (e.g., opal, carbonate, and aluminosilicate). Of all the analytical methods applied to marine waters and particles, at present only a small fraction can be systematically evaluated via comparison to reference materials that represent the appropriate natural concentrations and matrices.

Reference materials are homogeneous, stable substances whose properties are sufficiently established as to make them useful for calibrating analytical instruments or validating measurement techniques. High-quality reference materials not only provide essential support for large-scale research studies, but also ensure accuracy of long time-series measurements. In addition, with the advent of international oceanic data collection and distribution, reference materials allow comparison of data sets from different oceanographic studies. For example, since the development and widespread use of reference materials for oceanographic measurement of dissolved inorganic and organic carbon, it is possible to achieve excellent agreement among different cruises and groups as well as different times and locations. Reference materials have provided an important mechanism for oceanographers around the world to assess data quality and improve their methods. Comparative analyses of reference materials create benchmarks for cooperative, community-wide development of improved measurement methods, without risking the stagnation that can result from requiring the use of particular standard analytical procedures.

There are presently large programs in operation—and new projects being proposed—to measure chemical constituents in the ocean environment. These programs provide information on research areas such as: health and productivity of coastal oceans, sustainability of marine ecosystems, and predictability of climate change as well as other processes that affect the Earth's population on many levels. While some of the measurements made in these programs can be calibrated against reference materials, reference materials are lacking for many others, making it impossible to compare data over time and among groups.

These circumstances make the continued use of available chemical reference materials and the development of new reference materials essential. The National Research Council formed a committee in April 2001 to provide a comprehensive review of chemical reference material status and needs for those elements and compounds essential for investigating ocean processes. In addition, the committee was charged to provide advice on the elements and compounds requiring the development of reference materials and/or reference material improvement. Specifically, the committee was charged with the following tasks:

- compile from available sources a list of important oceanographic research questions that may benefit from chemical reference standards;
- create a comprehensive list of reference materials currently available for oceanographic studies;
- identify and prioritize the reference materials needed to study the identified research questions;
- determine for each priority analyte whether reference materials and/or analytic methods should be standardized; and
- identify the most appropriate approaches for the development and future production of reference materials for ocean sciences.

The Committee on Reference Materials for Ocean Science, sponsored by the National Science Foundation, held a workshop in September 2001 which brought together ocean chemists from multiple fields and agencies in the United States, Europe, and Canada to discuss the current and future needs for reference materials in ocean science. In addition, the committee surveyed the international ocean chemistry community with the aid of the American Society of Limnology and Oceanography to ascertain the community's concerns about needed reference material development. The interactions and the input from these groups allowed the committee to prioritize the need for chemical reference materials in seawater, suspended particulate, and sediment matrices. Furthermore, the committee offered recommendations to the oceanographic community and addressed social and educational issues. Box 1 lists the committee's recommendations for those reference materials that are needed to ensure the success of future global scale measurement programs as well as to further research in the ocean sciences on a broad number of fronts. Specific concentration information can be found in Chapter 6.

A limited number of reference materials have been explicitly developed for ocean science: salinity, ocean carbon dioxide, and dissolved organic carbon. Although salinity reference materials are available on a commercial basis from Ocean Scientific International Ltd. in the United Kingdom, the others are presently supported through grants from the National Science Foundation. The widespread use of such materials and their success in enhancing the scientific return on related studies is clear, and it is essential that such materials remain available. In addition, the National Institute of Standards and Technology presently prepares a number of standard reference materials that are of immediate use to the ocean science community. These include materials for ^{14}C (SRM 4990C) and ^3H (SRM 4361C) as well as ^{238}U, ^{234}U, ^{235}U (SRM 4321C), ^{230}Th (SRM 4342— presently out of stock), ^{226}Ra (SRM 4969), ^{228}Ra (SRM 4339B), ^{10}Be (SRM 4325), ocean sediment (SRM 4357), and river sediment (SRM 4350B). Again, it is important to assure the continued availability of these materials. In addition, a variety of *new* materials are needed.

Seawater reference materials are recommended for nutrients and for trace metals (especially iron). There is an urgent need for a certified reference material for nutrients. Completed global surveys already suffer from the lack of previously available standards, and the success of future surveys as well as the development of instruments capable of remote time-series measurements will rest on the availability and use of good nutrient reference materials. The reference materials for trace metals— though initially characterized only for the important micronutrient, iron— should ultimately be useful for the analysis of the other metals and some dissolved organic materials. The committee also recommends the devel-

Box 1
Recommended Reference Materials for Ocean Science

Materials Recommended for Continued Availability:

Currently available materials:
1. Standard Seawater (*Ocean Scientific International Ltd.*)
2. Reference Materials for Ocean CO_2 (*NSF via Dr. A. Dickson*)
3. Reference Materials for Dissolved Organic Carbon (*NSF via Dr. D. Hansell*)
4. Various Standard Reference Materials from NIST (see text)

Materials Recommended for Development:

Seawater-based reference materials:
5. One certified for the nutrient elements: nitrogen (as NO_3), phosphorus (as PO_4), and silicon (as $Si(OH)_4$).
6. One with concentrations of metals corresponding to oceanic deep water, certified for total iron concentration.
7. One with concentrations of metals corresponding to open-ocean surface water with an information value for total iron concentration.

Certified reference materials for radionuclides:
8. An acidic solution containing ^{238}U and ^{235}U with daughters in secular equilibrium through ^{226}Ra and ^{223}Ra.
9. An acidic solution containing ^{232}Th with daughters in secular equilibrium through ^{224}Ra.
10. An acidic solution containing ^{210}Pb with daughters in secular equilibrium through ^{210}Po.

*Solid matrix-based reference materials:**
11. Freeze-dried culture of the diatom *Thalassiosira pseudonana*
12. Freeze-dried culture of the dinoflagellate *Scrippsiella trochoidea*
13. Freeze-dried culture of the haptophyte *Emiliania huxleyi*
14. Open-ocean, carbonate-rich sediment
15. Open-ocean, silicate-rich sediment
16. Open-ocean, clay mineral-rich sediment
17. Coastal, carbonate-rich sediment
18. Coastal, silicate-rich sediment
19. Coastal, clay mineral-rich sediment
20. Deltaic sediment (that has not contacted seawater)

---

*Each of these solid reference materials should be certified for both inorganic and organic carbon concentrations, total nitrogen concentration, $\delta^{13}C$ of both the inorganic and the organic carbon components, and $\delta^{15}N$ for the total nitrogen component.

Box 2
**Additional Recommendations for Community
Participation**

1. The use of appropriate reference materials should be a key feature of the quality assurance/quality control structure in any future ocean science project involving chemical measurement. Reference materials use should be explicitly addressed in the project planning stages, proposals, and publications.

2. It is essential to develop and maintain a searchable, user-friendly database that ocean scientists can access to learn about those reference materials that are of particular interest to their research.

3. It is essential to encourage the presentation of short courses on the best way to use reference materials to ensure quality control of analytical measurements in conjunction with national meetings for ocean scientists.

4. It is important that round-robin exercises be organized using materials and analytes relevant to the ocean sciences and that laboratories be encouraged to participate, even at an early stage in their experience with the relevant analytical techniques.

5. It is important that the ocean science community be encouraged to investigate the various proposed matrix-based reference materials so as to establish their properties with consensus-based values for the concentrations of a variety of constituents.

6. Proposal and journal article reviewers need to be encouraged to question the analytical quality control of measurements made without the benefit of reference materials.

---

opment of three primary reference solutions for radionuclides that will be useful for a variety of ocean mixing and biogeochemical studies.

Particulate reference materials should be developed for three representative marine microalgae to further research on complex food webs, primary productivity, particulate fluxes, and ecological responses to climate change. The organisms proposed provide a wide-range of oceanographically relevant mineral, trace metal, and organic analytes and also represent three major marine matrices: opal, calcium carbonate, and organic matter. Furthermore, they may be used to prepare reference materials for the analysis of alkenones and photosynthetic pigments.

Sediment reference materials should be developed for both open-ocean and coastal areas. Open-ocean sediments should include carbonate-rich, silicate-rich, and clay mineral-rich types. Coastal sediments should be of the same types and should include a deltaic sediment that has not been in contact with seawater. Taken together with the algal-based materials, these sediment materials would represent a wide range of diagenetic states. The committee recommends that each of these solid materials (both algal- and sediment-based) be certified for both inorganic and organic carbon concentrations, total nitrogen concentration, $\delta^{13}C$ of both the inorganic and the organic carbon components, and $\delta^{15}N$ for the total nitrogen component.

In order for the ocean community to derive maximum benefit from the proposed reference materials, the committee made several additional recommendations (Box 2).

Thus, it is essential for the oceanographic community to maintain and promote the use of currently available reference materials while continuing the development of new ones. Furthermore, it is critical that the oceanographic community devotes energy to educating its members—from students and scientists to funding agencies and journal publishers—so as to ensure the quality of future chemical oceanographic observations.

1

Introduction

BACKGROUND TO THE STUDY

In this new millennium, during which human activity will bring about unprecedented change in the natural world, ocean scientists are called upon to address a variety of important challenges. The National Research Council (1998) chose to highlight three related research areas in a report, *Opportunities in Ocean Sciences: Challenges on the Horizon.* The focal areas selected were:

- Improving the health and productivity of coastal oceans,
- Sustaining ocean ecosystems for future generations, and
- Predicting climate variations over a human lifetime.

This report also notes that the oceans still remain too vastly undersampled in time and space to adequately address these global-scale long-term challenges.

In response to an earlier recognition of the growing need for a comprehensive characterization of ocean systems over time and space, the 1990s saw the start of various large research programs, such as the Joint Global Ocean Flux Study (JGOFS), designed to intensively study processes that cycle biologically active elements and compounds over key regions and time periods. New expedition-based sampling programs are presently in the planning stages. In addition, numerous observatories have been established over the last 10 to 15 years at coastal and open ocean sites to record the extensive time-series observations needed to recognize decadal-scale trends. These observatories include the Hawaii Ocean Time-series (HOT) and the Bermuda Atlantic Time Series (BATS) stations. Observations made at these sites have been particularly important in identifying changes in processes that regulate carbon cycling through large areas of the ocean (e.g., Karl, 1999). The specific scientific opportunities that can result from long-term observations over broad spatial scales have been highlighted in several recent reports (e.g., *Illuminating the Hidden Planet: The Future of Seafloor Observatory Science* (NRC, 2000); *Ocean Sciences at the New Millennium* (NSF, 2001); and *Discovering Earth's Final Frontier: A U.S. Strategy for Ocean Exploration* (President's Panel for Ocean Exploration, 2000).

In addition to the need for adequate sample coverage, comprehensive time-series studies of the global ocean require accurate, precise, and comparable chemical measurements of a daunting variety of dissolved and particulate constituents. Chemical characterizations have been a mainstay of oceanography for over a century. For example, precise measurements of total salt content are used to identify ocean waters of different origins and to predict how they will interleave and circulate. Analyses of dissolved nutrients such as nitrate (NO_3), phosphate (PO_4), and silicate ($Si(OH)_4$) indicate the potential for phytoplankton growth as well as the flow of food and energy to zooplankton and fish. Stable isotopes, trace elements, and organic molecules provide tags and tracers of events in the water column and sedimentary record, and radioactive isotopes serve as "clocks" to aid in determining the rates of the processes that produce these distribution patterns.

This work has relied on the skills and dedication of the individuals involved, who have typically carried out scientific studies of extremely high quality. However, even using modern techniques, chemical analyses of ocean waters and particles remains challenging given that they often involve low concentrations and complex sample matrices. Oceanographic measurements are often particularly prone to "matrix effects," where the target analyte is influenced by a variety of other sample components. In this time of global change, a central challenge for the oceanographic community is to synthesize data sets obtained at different times, by different laboratories using a variety of analytical techniques that are applied to contrasting sample matrices.

However, this is not a recent development. A 1971 report of the Marine Chemistry Panel of the National Academy of Sciences Committee on Oceanography wrote that:

> The rapid advance of marine science involves the participation of more and more people who are making more and more measurements.

This situation requires the development of better methods for managing the increased quantity and the quality of the data. For the former—the recording, storage, and digesting of the data—computer techniques are available and are becoming more common.

Quality control, however, needs more attention. For example, it has been reported that much of the chemical data produced by the International Indian Ocean Expedition is unusable because of doubts about its accuracy. Such reports are a perennial source of confusion in marine chemistry. Better calibration, universal standards, and interlaboratory comparison are essential if we are to continue our present field methods, in which independent investigators make measurements that are presumably comparable (NRC, 1971, pp. 54-55).

The need for oceanographic standards and interlaboratory comparisons was again reiterated more than 20 years later in the 1993 report of the Ocean Studies Board Committee on Oceanic Carbon:

The ability to make analytical measurements depends intimately on the availability of well-defined standards and calibrants. Many measurements of analytes in seawater (such as dissolved organic carbon and dissolved organic nitrogen) cannot be compared among laboratories because of the lack of appropriate reference materials and blanks for instrument calibration and testing. Intercomparison exercises are critical (NRC, 1993, p. 75).

In response to this clear and continually growing need, the current National Research Council Committee on Chemical Reference Materials for Ocean Science was empanelled in 2001 to:

- compile from available sources a list of important oceanographic research questions that may benefit from chemical reference standards;
- create a comprehensive list of reference materials currently available for oceanographic studies; (See Appendix E)
- identify and prioritize the reference materials needed to study the identified research questions;
- determine for each priority analyte whether reference materials and/or analytic methods should be standardized; and
- identify the most appropriate approaches for the development and future production of reference materials for ocean sciences.

Chemical Reference Materials

Reference samples and reference materials have served a critical role in analytical chemistry since its inception. The reliability of all analytical results is completely dependent on the availability of suitable reference materials, and now nearly all branches of analytical chemistry declare an

---

Box 1.1
Useful Definitions

Primary Standard: Standard that is designated or widely acknowledged as having the highest metrological qualities and whose value is accepted without reference to other standards of the same quality.

Secondary Standard: Standard whose value is assigned by comparison with a primary standard.

Reference Material: Material or substance whose property values are sufficiently homogeneous and well-established so as to be used for the calibration of an apparatus, the assessment of a measurement method, or for assigning values to materials. (Note: A reference material may be in the form of a pure or mixed gas, liquid, or solid. Examples include synthetic mixtures such as calibration solutions used in chemical analysis as well as materials based on natural environmental samples such as sediments.)

Certified Reference Material: Reference material, whose property values are certified by a procedure that establishes its traceability to an accurate realization of the unit in which the property values are expressed, and for which each certified value is accompanied by an uncertainty statement (certificate) at a specified level of confidence.

Traceability: Property of the result of a measurement or the value of a standard whereby it can be related to stated references, usually national or international standards, through an unbroken chain of comparisons, all having stated uncertainties.

---

urgent need for such standards (Zschunke, 2000). Specific terminology has evolved for different types of reference materials and is highlighted (Box 1.1) for use in later discussion.

There is often some confusion between the terms *standards* and *reference materials*. Primary standards represent the top-tier of chemical standards and, in principle, provide a means of establishing the traceability of analytical data to the SI measurement units (e.g., the kilogram, mole, meter, and second). A limited number of pure chemicals are recognized as primary standards (and thus can constitute certified reference materials). Most certified reference materials are not of themselves primary standards; rather, the property values assigned to them are traceable to primary standards where practical.

For the purpose of this report, the committee focused on reference materials for chemical composition measurements used in the ocean sciences. Although a number of these materials are simple mixtures or solutions, the majority of materials discussed are "compositional" or matrix reference materials. Such materials are based on "natural" substances (e.g., seawater, sediments, or biological materials such as phytoplankton), and offer an advantage over primary standards by providing a better match to sample composition. Thus they offer a tool to minimize matrix effects and to identify problems in the application of analytical methods to natural samples. If the reference material is a close match to the sample being analyzed, the measurement on the reference material will provide useful information on the quality of the laboratory's overall process. Conversely, analysts should be aware that matrix reference materials can contain constituents not present in their samples. If present at sufficiently high concentration, these constituents could prevent true matrix-matching between the reference material and the sample.

The highest quality reference materials are certified for the concentration values of the constituent(s) of interest, reflecting high confidence in the value's accuracy and the thorough investigation of all known or suspected sources of bias. Used appropriately (e.g., Roper et al., 2001; Zschunke, 2000), these *certified* values provide an effective means to ensure comparability (both among laboratories and over time). It is not always practical, however, to undertake the work required to produce a certified value; in many such cases, a value can be carefully determined, but insufficient information exists to assess the associated uncertainty. This information value is nevertheless of substantial interest to other users of the reference material. It can, for example, allow laboratories to compare results even though full traceability is impractical.

In this report, this distinction between *certified* and *information* values will be drawn on a number of occasions, usually to emphasize the cost of establishing the detailed uncertainty of analytical techniques used to certify materials relative to the perceived benefit of certification. In many situations, a particular analyte—though important scientifically—is studied by a limited number of researchers, in which case it is most practical to establish an information value for that particular reference material by consensus (provided the necessary conditions of stability and homogeneity are met).

BENEFITS OF CHEMICAL REFERENCE MATERIALS
TO OCEAN SCIENCE

It is impossible to have a discussion of the "quality" of chemical analyses made in the support of ocean science without a clear recognition that, in large part, this depends on the skills and dedication of the individuals involved. Indeed, almost all the scientific progress made to date in the ocean sciences has been achieved without the benefit of reference materials. Furthermore, reference materials are costly to produce—particularly if they are certified for a number of constituents—and it has not always been clear that this cost is repaid with significant added value.

However, it is also appropriate to consider the cost of *not* using reference materials. Reference materials provide the benefit of comparability—between results obtained at different times, in different places, by different people, and using contrasting methods. Almost every sub-discipline of ocean science research faces the need to understand complex dynamical processes that require large-scale, time-series data sets for effective study. When such data sets are acquired without using suitable reference materials, significant effort is expended later on efforts to adjust data to a common scale. In many cases this adjustment is impossible and the measurements have then, at best, limited value. Indeed, the costs may be far higher if erroneous management decisions are later based on such measurements.

The establishment of observatories constitutes a paradigm shift in the conduct of ocean sciences research—away from the traditional expeditionary mode—and coupled to this shift in observational modes is a need to ensure that an adequate infrastructure exists to support these observing systems and to provide the necessary quality assurance (NRC, 2000). In addition, the analytical tools needed for many large and small oceanographic studies are currently in a rudimentary form. Creating reference materials that can be exchanged between different laboratories will enable researchers to better understand their own techniques and the information they provide. Unfortunately, at present few certified reference materials exist for seawater properties, and many common marine matrices such as opal and carbonate bearing phytoplankton and sediments are not available as reference materials at all. Regardless, many research problems in the ocean sciences would benefit from the availability of suitable reference materials.

The recent *Future of Ocean Chemistry in the U.S. (FOCUS)* report (NSF, 2000) identified eight key themes for future chemical oceanographic research (Box 1.2). Clearly, oceanographic analyses over the coming decades will involve a wide variety of dissolved and particulate constituents from coastal ocean, open ocean, and seafloor settings, and thus a variety of reference materials will be required to represent these key sample matrices.

These eight research initiatives will require new strategies to ensure uniform analytical quality and comparability, and each of these areas could benefit to a greater or lesser extent from the availability of appropriate reference materials. Discussions at the Workshop (Islamorada, Florida in September 2001) held as part of this process helped the committee to clarify the ocean science community's perspectives on this, and

---

Box 1.2
Key Research Areas in Chemical Oceanography
(Based on FOCUS report; NSF, 2000)

(1) Major and minor plant nutrients—how they are transported to the euphotic zone, affect community structure, and how these processes are influenced by natural and anthropogenic changes.

(2) Land-sea exchange at the ocean margins.

(3) Organic matter assemblies at their molecular to supra-molecular scales, their reactivity, and interactions with other materials.

(4) Advective transport through sediments, coastal aquifers, and submarine ridge systems.

(5) Forecasting and characterization of anthropogenic changes in ocean chemistry.

(6) Air-sea exchange rates of gases that directly influence global ecosystems.

(7) Relationships among photosynthesis, internal cycling, and material export from the upper water column.

(8) Controls on the accumulation of sedimentary phases and their chemical and isotopic compositions.

---

provided a basis for future committee discussions during the preparation of this report.

However, the costs of developing new certified reference materials are substantial and new projects should not be entered into lightly. In choosing areas of ocean science to highlight by proposing new reference materials, the committee aimed to strike a balance between emphasizing a clear need for *certified reference materials* to assist in the quality control of long-term, multi-investigator observing programs, and identifying a parallel need for reproducible *reference materials* that will assist new communities of ocean scientists in improving their understanding of matrix problems in difficult and sophisticated analyses.

The committee thus recommends that a limited number of carefully chosen matrix reference materials be prepared (based on seawater, sediments, and algal materials) to respond to a significant portion of the current pressing needs. Such materials, certified for a limited number of key analytes, would represent stable homogeneous samples for a far wider range of additional constituents than is currently available. The committee recommends that these reference materials be investigated further by the scientific community to ascertain their usefulness for a wide range of constituents such as trace metals, organic compounds, and radioisotopes. Subsequently, consensus values for the concentrations of particular constituents should be assigned to many of these reference materials, further enhancing their usefulness.

REPORT STRUCTURE

Substantial discussion arose in deciding how to arrange this report as a consequence of the two-pronged nature of the problem. Should reference materials be thought of (and hence grouped for discussion) by analyte, or by matrix? Each of the natural materials occurring in the ocean system—seawater, biological materials, and sediments—provides a highly complex mixture of constituents. It is neither practical nor desirable to provide a complete analysis of all these constituents. Nevertheless, each matrix addresses particular scientific questions and provides particular challenges; the committee thus opted to arrange its discussion in terms of sample matrices.

Chapter 2 sets the stage with a brief statement of the present use of reference materials by the oceanographic community. A number of case studies have been included both to illustrate the value of such materials and to indicate the potential limitations of reference materials as a cure-all for oceanic measurements.

Chapter 3 discusses the analysis of seawater constituents and reiterates the scientific problems that necessitate a number of seawater-based reference materials.

Chapter 4 discusses the analysis of particulate materials such as sediments and biological materials. It outlines the current state of the science in some of these areas, and suggests that a limited number of particulate reference materials could play a significant role in moving these fields forward.

Chapter 5 provides specific suggestions for the production and distribution of reference materials for ocean science, noting some of the potential challenges and indicating possible strategies to avoid or mitigate these problems.

Chapter 6 provides a summary of the committee's principal recommendations that arise throughout the report and indicates those actions that the committee found to be of the utmost priority.

Appendix A contains biographical information on committee members. **Appendix B** contains the names and affiliations of the participants who attended the workshop in Islamorada, Florida in September 2001. **Appendix C** contains a glossary of technical terms used in this report. In addition, because this report relies on discussion of chemical elements, compounds, isotopes and radionuclides, a standard form of nomenclature was required. For this reason, where practical, the names of elements are spelled out throughout the report (e.g., nitrogen, carbon). The names of chemical substances are spelled out and the abbreviation given on their first mention (e.g., carbon dioxide [CO_2]). All isotopes and radionuclides are referred to by their standard chemical abbreviation (e.g., ^{14}C, ^{228}Ra). A list of the elements, compounds, radionuclides and acronyms referred to in this report, is also provided in **Appendix D** for clarification. **Appendix E** provides information about all currently available reference materials identified in this report, as well as sources for obtaining these (and other) reference materials.

2

Improving Chemical Oceanographic Data

Consistently comparable data collected on a global scale and over time are essential for the global monitoring programs and process studies described in the previous chapter. Individual investigators involved in mechanistic studies will also benefit from consistent and comparable analytical results. Although the oceanographic community has a strong record of emphasis on analytical quality control focused on precision, knowledge of the accuracy is often missing due to a lack of reference materials. Routine use of appropriate reference materials allows systematic analytical offsets to be recognized and related to known procedures or sample matrices. Availability of reference materials may actually encourage development and adaptation of new analytical methods as it will allow accurate and consistent comparisons. Bringing about community-wide use of reference materials on a routine basis is in part a matter of education. A key first step is to bring the issue to the attention of the oceanographic community, and to make the many advantages of comparative analysis clear. Example and peer pressure, especially from small groups of analysts that are already associated by common measurements, should also help. In addition, peer review of proposals and publications should also place more emphasis on the use of reference materials.

MATRIX DEPENDENCE OF REFERENCE MATERIALS

The subsequent chapters of this report are organized on the basis of matrix (e.g., seawater versus sediment), as opposed to analyte or measurement method. This choice is based on several considerations. Most importantly, seawater matrices must be sampled, processed, and analyzed differently than either marine sediments or particulate matter. Seawater is dominated by dissolved salts that comprise 3.5 percent of the fluid by weight and profoundly affect its physical properties (e.g., density, viscosity, and refractive index) as well as the types and rates of chemical reactions in solution. In contrast, marine sediment and other particulate matrices are dominated by minerals (typically calcium carbonates, opal or detrital aluminosilicates). Whereas seawater exhibits relatively constant compositions of the major ions (those present in amounts greater than 1 mg/kg), sediments are extremely variable in the types of minerals they contain and the chemical conditions encountered among and within deposits. Because it is seldom possible to quantitatively separate an analyte from its seawater or sediment matrix, the background ions and minerals must be carried through a major portion of most preparation and analytical procedures, where they can cause a broad range of complications and compromises.

A final rationale for the report's matrix-based organization is that seawater or sediment samples are often analyzed for different reasons by different types of oceanographers who may have little interest in another matrix. In general, analyses of seawater components address biological (e.g., photosynthesis/respiration), chemical (e.g., scavenging, reactions and mixing), and/or physical (e.g., circulation/mixing) processes imprinted over time scales of less than 1000 years, the mixing time of the global ocean. Particulate matter reflects biological and chemical processes in the water column corresponding to time periods significantly less than 100 years. Sediments are typically studied as temporal records of events occurring in the water column over time by sampling the accumulating deposit and can represent much longer time periods, extending up to millions of years. These contrasting oceanographic perspectives share a susceptibility to artifacts deriving from the unique matrices (whether seawater or sediment) from which chemical information is extracted.

HOW REFERENCE MATERIALS WORK

Routine use of reference materials can help tie together data sets, allowing us to answer key biogeochemical questions, even when the mechanisms of matrix effects and other analytical complications are not fully understood. This fundamental advantage can be illustrated by considering a scenario (Fig. 2.1) in which two different analytical methods for the same analyte are used to characterize an environment at two different times. In this example, the original data set (Fig. 2.1A) is ambiguous regarding whether the measured environmental variable has changed

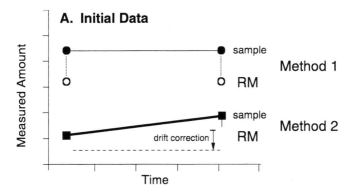

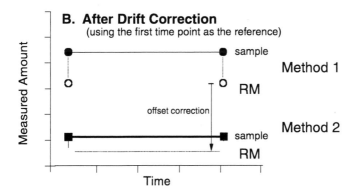

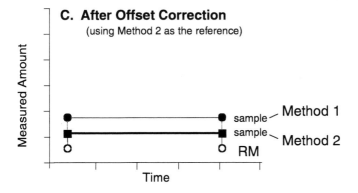

FIGURE 2.1 Comparison of two different analytical methods. (A) Original data set comparing the concentration of the analyte (sample) and a reference material (RM), (B) concentrations after correction for drift, and (C) concentrations after correction for reference material offset.

over time or whether the measurement method itself caused this shift. Method 1 indicates the concentration has remained constant with time, whereas Method 2 suggests it has changed. Parallel analysis of a reference material (whose composition remains unchanged over time) shows not only that Method 2 has drifted but also the extent of that shift. The drift-corrected data obtained by both methods now indicate that the concentration of the target analyte has not varied in the sample over time (Fig. 2.1B). Even after drift correction, the two methods measure very different analyte concentrations. Because the sample measurements are linked by simultaneous analyses of the reference material, however, the paired data show that the different concentrations measured by Methods 1 and 2 largely result from a systematic offset between these two procedures. Once this offset is removed (in this case, by adjusting all measurements made by Method 1 downward until the reference material values for both methods coincide), the sample measurements agree much more closely between methods (Fig. 2.1C). The overall result is that data obtained using the two methods can be confidently used together to constrain the value of the sample over time.

Reference materials can also provide a convenient means of identifying troublesome matrix effects and for probing the possible mechanisms for a given analytical method. For example, measurements of the same analyte in a variety of samples with different commonly-encountered matrices provide a powerful means of identifying unusually large offsets between different analytical methods. Such sample-dependent inconsistencies serve as a warning that at least one of the compared methods is unusually sensitive to some characteristic of the sample matrix. Because sample matrix effects can be extremely complex and analyte-specific, it is highly advantageous for the oceanographic community to investigate these complications using a small number of widely-shared reference materials. In turn, these reference materials should represent the major matrix types encountered in natural samples, and they should contain a wide array of compounds in various physical forms and stages of degradation.

LIMITATIONS OF REFERENCE MATERIALS

The availability of reference materials or standards will not solve all the analytical problems faced by the marine community. In addition to using reference materials, the use of agreed-on common collection and analytical methods can also improve the chemical data being collected by oceanographers. Standardization of these methods minimizes the variability that may result from differences in laboratory procedure. A major disadvantage of method standardization, however, is that it can discourage the development of improved analytical methods. In the specific case of oceanographic studies, where accurate analyses of major elements in many commonly encountered matrices are not yet routine, it is therefore premature to standardize measurement methods for most analytes. Nevertheless, even in the absence of reference materials, there are well-developed approaches to the quality control of measurements. For example, careful attention needs to be paid to (1) blanks, (2) calibration, (3) replicate analyses, and (4) where practical, the use of "spike recovery" to assess the quality of analyses.

Different sampling methods often result in collection of different components of the element or compound of interest. Most seawater samples are collected in bottles, filtered to remove particles and analyzed directly or after preconcentration of minor components. Particulate matter from the water column is collected by filtration or with sediment traps. Sediment samples are collected in cores that recover intact chronological sequences and are commonly subsampled, dried and ground to a powder. Pore waters are extracted from sediments by squeezing or suction, dialysis, and centrifugation.

One example of a collection issue that cannot be solved using reference materials is particulate matter filtration. This widespread pretreatment has led to the operational definition that "particulate" material is retained by a filter (commonly between 0.2 and 0.7 μm pore-size), whereas "dissolved" material passes through the filter. Although this definition was arbitrarily based on the minimum pore size filters available to early oceanographers, it allows most living organisms (except smaller bacteria and viruses) and particles of sufficient size to sink to be concentrated and separated. The choice of filters and even the filtering technique can greatly affect the composition of the sample that is analyzed. Studies of colloidal matter and its importance in aqueous processes are still at an early stage, but will have additional impacts on the separation into "dissolved" and "particulate" materials.

A current example of these filtering artifacts is the measurement of particulate organic carbon (POC) and chlorophyll concentrations in the water column. Because organisms vary in size and fragility, the choice of filter can determine the types of organisms that pass through or are retained, and thus the measured concentrations of both the particulate and dissolved species of interest, while filtration pressure can influence whether the organism is ruptured during the process. The size of the water sample can affect the analysis for materials that adsorb onto the filter; for example, dissolved organic carbon (DOC) can adsorb onto filters and may bias particulate organic carbon analysis in very small samples. Standardization of collection methods allows many laboratories to measure analytes in the same particulate component, but certain questions may require measurement of a different component. The best way to resolve this type of problem is for authors to report in detail their collection methods so that later investigators will know what component was being measured and can draw their own conclusions. Journals should be encouraged to recognize the need for publication of such information.

BENEFITS OF REFERENCE MATERIALS

The following boxes present four case studies to illustrate how the introduction of reference materials has decreased the uncertainty of the chemical oceanographic measurement of salinity (Box 2.1), DOC (Box 2.2), and dissolved inorganic carbon (DIC) (Box 2.3). Box 2.4 illustrates the acute need for pigment reference materials, which are currently unavailable.

**Box 2.1
Case Study: Salinity (*S*)**

Early chemists found that the relative composition of the major components of seawater were constant (Marcet, 1819), meaning that the measurement of one component of seawater could yield values for many of the remaining elements (Millero, 1996). It also meant that one could determine the total dissolved salts or salinity (*S*, parts per thousand, ‰) in seawater from the measurement of one element. The element of choice was chlorine. Knudsen and co-workers defined chlorinity (*Cl*) as "the chlorine equivalent of the total halide concentration in parts per thousand by weight, measured by titration with silver nitrate ($AgNO_3$) solution" (Forch et al., 1902). The salinity determined from the evaporation of seawater ("Salinity is to be defined to be the weight of inorganic salts in one kilogram of sea water, when all the bromides and iodides are replaced by an equivalent amount of chlorides, and all carbonates are replaced by an equivalent quantity of oxides.") was directly related to chlorinity *Cl* (‰) as measured above. For over 60 years researchers used that relationship, as expressed in the following equation, to determine the salinity and subsequent density of ocean waters:

$$S\ (‰) = 0.03 + 1.805\ Cl\ (‰).$$

Early scientists recognized that standards were needed to determine reliable values of the chlorinity and salinity of seawater. The IAPSO Standard Sea Water Service (originally based in Copenhagen) collected and distributed seawater from the North Atlantic with a known, measured chlorinity. This sample was supplied to oceanographers to standardize the $AgNO_3$ solutions used to determine chlorinity in various laboratories.

In the 1960s more and more oceanographers were trying to determine the salinity of seawater using conductivity measurements. Attempts were made to develop correlations of the conductivity ratio of ocean samples with respect to seawater with a known chlorinity. The values of chlorinity determined in this manner were then converted to salinity using the Knudsen equation given above. Studies revealed, however, that the conductivity ratio of standard seawater varied for waters of the same salinity or chlorinity, which led to the development of the practical salinity scale that related the conductivity ratio to a fixed mass of potassium chloride (KCl) in a given mass of water. The seawater used to fix the scale had a salinity equal to 35 (*S* = 35.000) that was consistent with the Knudsen equation (*Cl* = 19.374) and thus with earlier measurements. Conductivity derived salinity is thus no longer directly connected to the chlorinity of the seawater; standards now have a specified conductivity ratio that can be used to calibrate salinometers with a sample of known salinity. The conductivity standard has made it possible for ocean chemists worldwide to make a routine determination of seawater salinity that can be compared with great precision and accuracy.

Box 2.2
Case Study: Dissolved Organic Carbon (DOC)

The measurement of DOC in seawater has long been a challenge to oceanographers and marine scientists (Hansell and Carlson, 2002). Indeed, ever since Natterer (1892) and Putter (1909) made the first measurements of DOC using different methods, there has been a continuing controversy over the levels of organic carbon in seawater. Resolution of this issue appeared to be imminent in the late 1970s following the publication of several comparison studies (e.g., MacKinnon, 1978; Gershey et al., 1979). In 1988, Sugimura and Suzuki reported dramatically higher DOC levels along with a remarkable correlation of these new observations with apparent oxygen utilization (AOU). The findings sparked a re-evaluation of the various methods used to determine DOC and opened a discussion regarding the validity of widely held notions regarding the nature, concentration, and distribution of dissolved organic matter in the ocean (Williams and Druffel, 1988). It would take another decade before the issue was settled.

An early attempt to resolve the discrepancy between the high values of Sugimura and Suzuki (1988) and more traditional analyses failed to reach a definitive conclusion (Williams, 1992). The start of the Joint Global Ocean Flux Study (JGOFS) field program with the North Atlantic Bloom Experiment in 1989 put additional pressure on the various groups to resolve this issue quickly. The National Science Foundation (NSF) and the National Oceanic and Atmospheric Administration (NOAA) funded a workshop held in Seattle in July 1991 to resolve the issue.

Prior to the Seattle Workshop, several batches of seawater from Hawaii were distributed to attendees for analysis. It became immediately clear to workshop participants that the key to making valid comparisons was both a common reference material and a uniform blank solution (Hedges et al., 1993; Sharp, 1993). The primary source of discrepancy among analysts was poor blank control, not oxidative capacity.

During the JGOFS EqPac cruises (1992) a smaller group was supplied with intercomparison samples and a uniform blank solution (Sharp et al., 1995). This effort led to a broad community intercalibration exercise for the accurate determination of DOC concentrations (Fig. 2.2). Conducted over the past five years, 62 laboratories from 17 nations participated in these intercalibration exercises. Most groups of analysts showed comparability better than 10 percent, while the most experienced analysts, sharing reference materials, showed reproducibility at the 2 percent level (Sharp et al., 2002). At the Seattle Workshop, DOC concentrations had varied by as much as a factor of three, and there was little common agreement regarding background levels for DOC in deep-ocean waters. Dennis Hansell has continued the production of a seawater DOC reference material and blank water reference material that is now being distributed internationally at minimal cost to the investigators (Hansell, 2001). Continued support of this program (currently funded by NSF) is essential to maintaining the high quality of research results; omission of the use of these standards is slowly becoming grounds for rejection of research papers by some peer-reviewed journals.

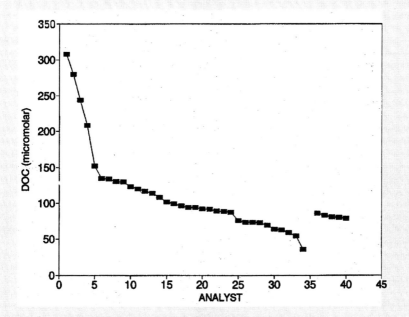

FIGURE 2.2 Example of the use of reference materials on the agreement between analysts. Analysts 1-34 represent data from all analyses of the Pacific Ocean surface sample from the Seattle Workshop (Hedges et al., 1993). These analyses were purposely performed without common reference materials or uniform blank correction in order to assess the state-of-the-art. Analysts 36-40 represent a select group of JGOFS investigators analyzing the EqPac inter-laboratory comparison sample using a common blank water and reference standard. The improvement in precision is immediately apparent.

Box 2.3
Case Study: Dissolved Inorganic Carbon (DIC)

High quality measurements of ocean CO_2 have been an integral part of the JGOFS and World Ocean Circulation Experiment (WOCE) programs. Despite their importance for understanding the oceanic carbon cycle, past measurements made by different groups were rarely comparable. A significant contribution of the JGOFS effort has been to produce and distribute reference materials for oceanic measurements. With funding from the NSF and Department of Energy (DOE) since 1989, Andrew Dickson's laboratory at the Scripps Institution of Oceanography has prepared over 50 separate batches of DIC reference material and has distributed more than 25,000 bottles of this material to scientists in over 25 countries.

During a recent expedition in the Indian Ocean, as part of the WOCE Hydrographic Program, members of the U.S. DOE CO_2 Survey Science Team used these reference materials extensively for the quality control of measurements of total DIC and total alkalinity. Two manuscripts detailing the results of these measurements on reference materials were published describing the CO_2 measurements made on that expedition and how reference materials were used to assess the overall data quality (Johnson et al., 1998; Millero et al., 1998).

Another indication that the use of reference materials has improved oceanographic data quality can be seen by examining the degree of agreement between measurements for deep water masses obtained where two separate cruises intersect. Lamb et al. (2002) examined this in detail for cruises in the Pacific Ocean and showed that the measurements of total DIC (for cruises where reference materials were available) typically agreed to within 2 µmol/kg (Fig. 2.3). This is in sharp contrast to the required adjustments to previous oceanic carbon data sets over the years.

A comparison of the concentration of ^{13}C[1] over a time span of 10 to 20 years can be used to calculate the uptake rate of anthropogenic CO_2 (Quay et al., 1992). Although extensive sampling occurred during the GEOSECS program in the 1970s, it has been impossible to use the ^{13}C data because of concerns about their reliability (Kroopnik, 1985). The lack of ^{13}C standards during the 1970s therefore has rendered much of the historical isotopic data useless.

---

[1]Measurements of ^{13}C and ^{14}C are reported as $\delta^{13}C$ and $\Delta^{14}C$ values. Both ^{13}C and ^{14}C are present in very small quantities relative to ^{12}C ($1\text{-}10^{-10}\%$) and the $\delta^{13}C$ and $\Delta^{14}C$ notation permits one to see small differences easily. (For further detail, see Appendix D.)

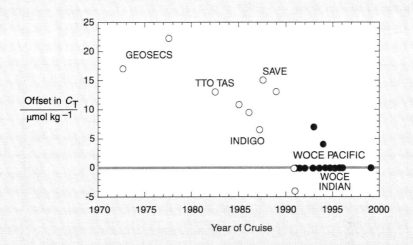

FIGURE 2.3 Estimated data offsets for total DIC measurements on various cruises (Gruber et al., 1996; Gruber, 1998; Sabine et al., 1999; Lamb et al., 2002). Certified reference materials for CO_2 became available in 1991 and were used on cruises represented by filled circles. (Note: The period 1991-2000 includes 35 cruises, most with no offset.) Labels refer to ocean-ographic research programs. Printed with permission of A. Dickson.

Other cruises beginning in the 1970s on which $\delta^{13}C$ was measured provided data that were used to calculate CO_2 uptake rates (Quay et al., 1992). Since no reference materials were available during that time, these calculations relied on comparison of measurements made on samples from depths greater than 2000 meters with the assumption that the $\delta^{13}C$ content below this depth would not change over the decadal time scale. Originally it appeared that there was agreement between the $\delta^{13}C$ from the 1970s values and those measured in the late 1980s and early 1990s.

It is possible to use the increased sampling and greater geographic coverage provided by the JGOFS and WOCE $\delta^{13}C$ programs in the 1990s (McNichol et al., 2000; Quay et al., submitted), to make more deep water comparisons at stations that are geographically closer. Using this new data it is possible to show that the deep water data from the 1970s is offset from that of the 1990s by 0.14 to 0.20 ‰ (Lerperger et al., 2000; Quay et al., submitted). Applying this offset to the data would significantly alter the calculated uptake rates. There is anecdotal evidence, however, that surface water samples were collected and analyzed differently from the deep water samples, making it unclear whether the observed offset should be subtracted from the surface water data.

Box 2.4
Case Study: Pigments

Pigment intercalibration exercises have been performed in support of NSF's JGOFS program (Latasa et al., 1996) and for NASA's Sea-viewing Wide Field-of-view Sensor (SeaWiFS) project (Hooker et al., 2000). For the first exercise, pure individual and mixed pigment standards were distributed to eight JGOFS pigment laboratories. Results from three separate intercalibration exercises documented a better agreement for spectrophotometric analyses than for high pressure liquid chromatography (HPLC) analyses. For the spectrophotometric comparisons, 90 percent of the pigments analyzed by participant laboratories were within six percent of mean consensus values. By comparison, 65 and 85 percent of the laboratories agreed to within 10 and 20 percent, respectively, on HPLC analyses. Chlorophyll absorption measurements obtained with diode array-type spectrophotometers were six to nine percent lower than those obtained with monochromator-type spectrophotometers. Furthermore it was determined that the use of HPLC methods incapable of separating monovinyl chlorophyll a from divinyl chlorophyll a can result in the overestimation of total chlorophyll a concentration in *Prochlorococcus*-dominated oceanic waters by 15 to 25 percent. A simple dichromatic approach was suggested for eliminating this variable source of error, which is caused by co-elution of these structurally-related pigments (Latasa et al., 1996).

For the second exercise, marine particulate matter collected at twelve stations in the Mediterranean Sea was distributed in triplicate to each of four SeaWiFS validation laboratories. Each laboratory used a different HPLC method for determining the concentrations of 15 different pigments or pigment associations. The four methods for determining concentrations of total chlorophyll a (i.e., monovinyl plus divinyl chlorophyll a) and the full set of pigments agreed within 7.9 percent and 19.1 percent respectively. In addition, accuracy was reduced by approximately 12.2 percent for every order-of-magnitude decrease in pigment concentration.

Standards and samples prepared for these two intercomparisons were consumed during those studies and were not available for JGOFS Antarctic Environment and Southern Ocean Process Study (AESOPS) cruises. Comparisons of chlorophyll a concentrations measured by fluorometry (Chl_F) and by HPLC (Chl_H) during AESOPS indicated large discrepancies for three of the four Polar Front cruises. The large overestimates in Chl_F observed may have been caused by interference associated with chlorophyllide a and chlorophyll c (Trees et al., 2000). The measurement disparities and probable causes described above required several months of additional data analysis and interpretation. The availability of pigment reference materials during the AESOPS program would have minimized the time required to sort out the observed discrepancies through the identification (or elimination) of potential analytical problems (e.g., operator error, accessory pigment interference, poor extraction efficiencies, calibration problems, etc.) during data collection. Standardization of pigment methodology for use with the HPLC technique would have resulted in more accurate and precise data.

3

Seawater

NUTRIENTS

Measurements of the major elements composing organic matter in ocean environments are among the longest established, most fundamental, and broadly informative analyses in the marine sciences. This tradition is based on the fact that six elements (carbon, hydrogen, oxygen, nitrogen, sulfur, and phosphorus)[1] make up essentially the entire mass of marine organic matter and largely govern the processes by which these elements are cycled in the ocean. One of the most widely used empirical relationships in oceanography is the Redfield-Ketchum-Richards equation (Redfield et al., 1963):

$$106 \ CO_2 + 16 \ HNO_3 + H_3PO_4 + 122 \ H_2O \rightarrow$$
$$(CH_2O)_{106}(NH_3)_{16}(H_3PO_4) + 138 \ O_2$$

This equation describes the ratios with which inorganic nutrients dissolved in seawater are converted by photosynthesis into the biomass of "average marine plankton" and oxygen gas O_2. The opposite of this reaction is respiration, or the remineralization process by which organic matter is enzymatically oxidized back to inorganic nutrients and water. The atomic ratios (stoichiometry) of this reaction were established by

---

[1]Symbols for these elements are as follows: C, H, O, N, S, P.

Redfield (1934), who analyzed the major elemental content of many samples of mixed plankton (phytoplankton and zooplankton) caught in nets towed through the surface ocean. They compared the carbon, nitrogen, and phosphorus composition of these collections to concentration profiles of dissolved inorganic carbon (DIC), NO_3, and PO_4 throughout the water column. This pioneering research demonstrated that these three elements are continually redistributed in the ocean by selective removal into plankton cells and their remains (i.e., fecal pellets), which are then efficiently respired as they sink through the marine water column.

As a result of continual nutrient removal and regeneration, the surface waters of density-stratified temperate oceans become so depleted in NO_3 and PO_4 that they limit the rate of photosynthesis, and hence the total amount of life that can be sustained through the food web. The extent of depletion in dissolved O_2 concentrations in the deep ocean is an additional useful chemical indicator of the total amount of respiration and oxidation that has occurred since downwelling from the surface. Using the Redfield-Ketchum-Richards equation, the measured apparent oxygen utilization (AOU) can be used to estimate the amounts of CO_2, NO_3, and PO_4 generated by heterotrophic activity within the same water parcel and, by difference, the concentrations of NO_3 and PO_4 that were initially present prior to downwelling. Such preformed nutrient concentrations can be useful tracers of the origins and mixing ratios of seawater that has entered the deep ocean from different surface regions (Broecker, 1974; Broecker and Peng, 1982). Measurements of the elemental composition of marine plankton thus have provided a basis for estimating the source, age, and life-sustaining potential of ocean waters for over 50 years.

In addition to the previous applications, the NO_3 to PO_4 ratio (N:P), for example, can be used to examine the cycling of nutrients in surface and deep waters (Wu et al., 2000). The N:P ratios at the Bermuda station (BATS) and at the Hawaii station (HOT) in surface waters are quite different (Fig. 3.1). Near Bermuda, the values of N:P range from over 30 at the surface and decrease steadily to approximately 17 below 600 m, while near Hawaii, they are nearly zero in the surface waters and increase with depth to a value of around 15 by 400 m. These results suggest that the balance between production, export, and mineralization of organic matter differs at the two sites. Furthermore, the measurement of nutrients over a 10-year period at the BATS station (Fig. 3.2) shows large variations that can be related to an extended period with a positive North Atlantic Oscillation index resulting in warmer-than-average sea surface temperatures and a reduction in the extent of convective overturn. Hence, less nitrate is mixed to the surface (Michaels et al., 2001). Future long-term studies of this type could benefit from the use of nutrient reference mate-

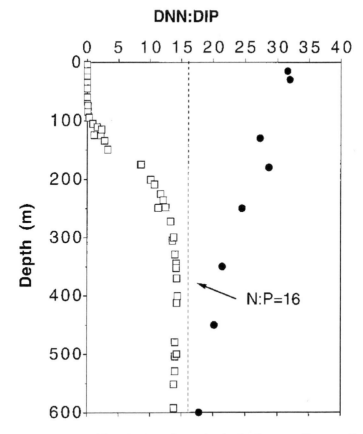

FIGURE 3.1 Vertical profiles of N:P molar ratios in the Sargasso Sea near Bermuda (31.67°N, 64.17°W) (•) and in the Pacific near Hawaii (HOT US JGOFS Web site at hahana.soest.hawaii.edu) (□). (Wu et al., 2000).

rials ensuring that measurements can be compared at different laboratories over long time periods.

The urgent need for nutrient standards was demonstrated during the recently completed World Ocean Circulation Experiment (WOCE) and Joint Global Ocean Flux Study (JGOFS) measurements which were made by different laboratories. The internal consistency of the nutrient data was evaluated by comparing measurements made in deep water (depth over 3500 m) at nearby stations on different cruises. If one assumes that nutrient concentrations in deep water at the same location should not change over decadal time scales, the observed systematic nutrient differences should represent an estimate of inter-laboratory precision. Offsets were found, however, among the results of different laboratories indicating inconsistencies in the preparation of the calibration standards. Mean offsets were 0.5 to 0.7 μM for NO_3, 0.05 μM for PO_4, and 2 μM for $Si(OH)_4$ (Mordy et al., 1999; Ross et al., 1999; Zhang et al., 2000). These offsets are equivalent to 1.5 to 2.0 percent of the respective deep-water nutrient concentrations. Corrections were frequently applied to the raw data to improve the internal consistency of the overall data sets, although many nutrient chemists have pointed out that such offset corrections are problematic (Gordon et al., 1999; Zhang et al., 2000). Without improvements to the accuracy and precision of nutrient measurements, it will be impossible to detect small changes in nutrient concentrations that may be important for understanding the oceanic carbon cycle.

Since certified reference materials for seawater nutrient analysis are currently unavailable, individual laboratories must prepare their own standard solutions for instrument calibration. Standard stock solutions are prepared at high concentrations (mM) so that they can be used for months without significant alterations in concentration. Working low-concentration standard solutions are unstable and need to be prepared daily by diluting stock solutions with distilled water or low-nutrient seawater. In this case, the accuracy of nutrient analysis at a given laboratory is highly dependent upon the accuracy of the daily preparation of the calibration solutions.

In the 1970s, the Sagami Chemical Research Center in Japan provided nutrient reference material for the Cooperative Study of the Kuroshio Current (the so-called CSK standards). These solutions were not prepared in seawater, which limits their general utility (see below), however they are still distributed and widely used as a common reference. French and British scientists have conducted some studies on nutrient reference material (Aminot and Keroul, 1991, 1996; Zhang et al., 1999) with limited success.

Matrix effects in the analysis of nutrients in seawater are caused by differences in background electrolyte composition and concentration (salinity) between the standard solutions and samples. This effect causes several methodological difficulties. First, the effect of ionic strength on the kinetics of colorimetric reactions results in color intensity changes with matrix composition and electrolyte concentration. In practice, analytical sensitivity depends upon the actual sample matrix. This effect is most serious in silicate analysis using the molybdenum blue method. Second, matrix differences can also cause refractive index interference in automated continuous flow analysis, the most popular technique for routine nutrient measurement. To deal with these matrix effects, seawater of standard salinity (approximately 35) should be used as the matrix for nutrient reference materials.

TRACE METALS

Measurement of trace metal concentrations can provide fundamental insights into marine geochemical processes. Many metals are important micronutrients in seawater and can play a significant role in upper ocean biogeochemistry and carbon cycling. Under certain conditions, elevated concentrations of metals associated with human activities can have negative impacts on marine ecosystems.

Comprehensive understanding of trace metal geochemistry eluded oceanographers for decades because of the difficulty of measuring metals at very low concentrations, particularly in open ocean waters. Many of the most important metals are also highly prone to contamination. Over the last 25 years, most of these obstacles have been overcome by a relatively small number of research groups. Now, with an increasing number of investigators becoming interested in high quality trace metal measurements, there is a greater need for reference seawater with metals at realistic (i.e., low) concentrations for researchers to assess their methodologies.

In existing reference materials the concentrations of several key analytes, including iron, are too high to be useful for scientists making open ocean analysis. For instance, the concentration of iron in seawater standards provided by NRC-Canada is about 100 times greater than expected in surface ocean waters (Table 3.1).

TABLE 3.1 Comparison of Metal Concentrations in an Existing Reference Seawater (NASS-5, from NRC-Canada) and in Oceanic Seawater

Metal	NASS-5 nM	Pacific Surface Water nM	Pacific Deep Water nM	Reference
Cadmium (Cd)	0.2	0.002	0.8	Bruland, 1980
Cobalt (Co)	0.2	0.01	0.08	Martin and Gordon, 1988
Copper (Cu)	5	0.6	5	Bruland, 1980
Iron (Fe)	3.8	0.06	0.7	Martin et al., 1989
Lead (Pb)	0.04	0.04	0.05	Wu and Boyle, 1997
Zinc (Zn)	1.6	0.08	8	Bruland, 1980

The highest priority metal that requires a reference material is iron, which limits primary production in approximately 40 percent of the world's oceans. Iron levels are extremely low (subnanomolar) in ocean surface waters of these regions, although there is disagreement among laboratories regarding the actual concentration. Difficulties in analyzing this highly contamination-prone element, coupled with an increasing recognition that iron concentrations exhibit considerable spatial and temporal variability, make the development of low-level reference materials for iron a high priority.

There are other metals for which compelling cases can be made to produce contamination-free oceanic reference seawater. These include other bioactive metals (e.g., zinc, cobalt, cadmium, and copper), tracers of anthropogenic contamination (e.g., lead, Box 3.1), and non-bioactive metals used as tracers of geochemical and physical processes (e.g., aluminum).

The primary and immediate need is for a trace metal reference material, but a certified reference material would provide even greater benefits. A technique based on isotope dilution with detection by inductively-coupled plasma mass spectrometry (ICP-MS) (Wu and Boyle, 1998) most clearly meets the traceability criteria required for a certified reference material. Although useful for iron and several other metals, isotope dilution is not possible for monoisotopic elements like cobalt, so other techniques must also be used. Indeed, it is advisable that several techniques be used to certify a trace metal reference material.

Box 3.1
History of Metal Analysis in Sea Water

In the 1960s much of the interest in metals centered on their role as pollutants. Considerable progress was made in characterizing the distribution of bomb-derived radionuclides, but little progress was made for stable metals. One of the first success stories was obtained by Schaule and Patterson who documented the marked effect of leaded gasoline on the distribution of lead in the upper water column (Schaule and Patterson, 1981). Wu and Boyle (1997) subsequently documented the removal of lead from the upper water column by scavenging processes, coupled with a dramatic decrease in deposition associated with the phasing out of leaded gasoline (Fig. 3.3). Such data, where decade-scale trends are superimposed over strong intra-annual variability, illustrate the importance of accuracy and precision in trace metal analysis. Widely available reference materials can help provide the analytical continuity over time and between laboratories required for such challenging trace analyses.

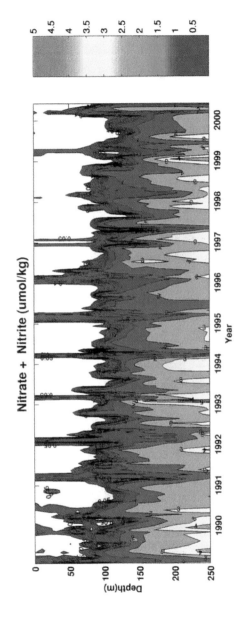

FIGURE 3.2 The variation of nitrate (+ nitrite) concentrations in surface waters at the Atlantic (BATS) time series study site (Michaels et al., 2001).

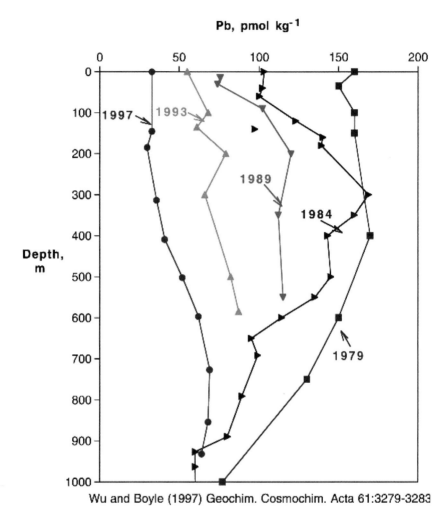

Lead Profiles near Bermuda, 1979-1997

Pb, pmol kg^{-1}

1997

1993

1989

1984

1979

Depth, m

Wu and Boyle (1997) Geochim. Cosmochim. Acta 61:3279-3283

FIGURE 3.3 Depth profiles in the North Atlantic showing the decrease of lead since the introduction of unleaded gasoline (Wu and Boyle, 1997).

Interference in Trace Metal Measurements

The major anions and cations in seawater have a significant influence on most analytical protocols used to determine trace metals at low concentrations, so production of reference materials in seawater is absolutely essential. The major ions interfere strongly with metal analysis using graphite furnace atomic absorption spectroscopy (GFAAS) and inductively coupled plasma mass spectroscopy (ICP-MS) and must be eliminated. Consequently, preconcentration techniques used to lower detection limits must also exclude these elements. Techniques based on solvent extraction of hydrophobic chelates and column preconcentration using Chelex 100 achieve these objectives and have been widely used with GFAAS.

The advent of ICP-MS has revolutionized seawater analysis for several reasons. First, while it is sensitive to interferences from sodium and chlorine, magnesium matrix effects are modest and can be corrected for. Now, many metals can be determined by co-precipitation from seawater with magnesium hydroxide ($Mg(OH)_2$) and re-dissolution into nitric acid. This technique is simple, and avoids contaminants associated with synthetic chelates. Second, because of its high sensitivity, pre-concentration factors of 100-1000 are no longer needed, and reliable measurements can be generated from 1-10 ml of seawater. Third, because individual isotopes are measured from the mass spectra, the technique of isotope dilution can be used to determine metal concentrations. This approach, described in detail in Wu and Boyle (1997), involves spiking a sample with a stable isotope that is affected in the same way as the target analyte by matrix effects, ICP-MS sensitivity fluctuations, and variability in the recovery efficiency of the precipitate. This is a huge advantage in artifact-prone matrices like seawater.

Dissolved organic matter (DOM) can complicate the analysis of metals such as iron and copper, which form strong organic complexes that render them less reactive. In oceanic waters, it is generally thought that prolonged acidification at pH 2.3 is sufficient to dissociate metals from natural organic ligands. Some workers also irradiate waters with ultraviolet light to destroy organic complexes, but a rigorous comparison of these approaches has not been carried out. It may be necessary to use the simplest procedure (acidification only) to minimize the chance of sample contamination. Dissolved organic matter also has surfactant properties that can interfere with the determination of metals by electrochemical techniques. Surfactants can also be eliminated by UV irradiation, and most electrochemists irradiate prior to determining total metal concentrations.

In coastal waters containing very high dissolved organic carbon

(DOC), acidification alone may not be sufficient to render all of a given metal reactive towards the reagents used for analysis. This may be an issue in important regions like major estuaries and waters near river deltas. Progress in this aspect of analysis may be greatly facilitated by a high-DOC reference seawater, though the lability of metals in such a sample may change with prolonged storage under acidified conditions.

Recommended Reference Materials

The principal need for trace metal reference materials is for a large volume of contamination-free seawater that can be disseminated to a large number of investigators. The highest priority for a trace metal reference material is iron. Two seawater-based reference materials are recommended: one from deep water having a relatively "high" iron concentration (approximately 0.6 nM), and one from an iron-depleted near-surface water (approximately 15 m) where the concentration is at least 10 times lower. For many analysts, the low-concentration sample would be useful for establishing blanks, and the high-concentration sample would be useful for establishing precision.

The same samples could also be used to analyze metals such as zinc, manganese, copper, molybdenum, cobalt, vanadium, lead, aluminum, cadmium, and the rare earth elements. Some of these elements have nutrient-like distributions, so surface and deep water samples would have significantly different levels. Surface water concentrations of critical elements like zinc, however, are not always low in regions where iron is low (e.g., in the Southern Ocean). For this reason, collection locations for trace element reference materials must be carefully selected. Participants at the Florida workshop in September 2001 (Appendix B), suggested the southeast tropical Pacific as a region where iron and zinc might be at low concentrations in surface waters. Manganese and aluminum concentrations may be comparable or higher in the surface sample than in the deep sample, but either sample would be valuable as a reference material for these elements.

Two additional reference materials should also be considered, one for coastal waters and one for the Great Lakes. A standard for dissolved iron and other metals in coastal water containing high concentrations of DOM would be useful in addressing the matrix effects associated with DOM described above. Interactions with organic matter can make metals more difficult to extract from seawater and cause poor recoveries (even for laboratories experienced in the analysis of seawater using clean techniques). One potential problem that will require monitoring is any change in the reactivity of dissolved iron in a very high DOM sample during prolonged storage. The Great Lakes are considered "inland seas" by the

National Science Foundation (NSF), and the Chemical Oceanography program invests substantial resources in that area. The upper water column in the center of Lake Superior might be a good place to collect water for preparation of a freshwater trace metal reference material.

A first attempt at the collection of a low-iron oceanic surface water reference material—part of a Scientific Committee for Oceanic Research (SCOR)-sponsored inter-comparison that was co-funded by the European Union—was only partially successful. The reference material had a higher iron concentration than expected after it had passed through the preparation stage, illustrating the difficulty associated with avoiding contamination of large water samples on board an iron ship. This "IRONAGES" project provided a valuable learning experience. This attempt cost approximately 100,000 Euros, for coordination of homogeneity tests, and distribution of the reference material. As of 2002, the reference material was being used in an international laboratory intercomparison.

The collection of a reference material for trace metals could complement collection of samples for other priority analytes including DOM and DOC. Seawater used for DOM analysis can also be collected in Teflon®, for example. Although samples for DOM are not normally stored in plastic bottles (as the iron reference material would be), releases of plasticizers are minimal and are not expected to interfere with the detection of individual organic compounds such as amino acids and sugars.

It is strongly recommended that sample collection be linked to intercomparisons of field-based methodologies by different groups of investigators. This would help justify the expense of the required ship time, and satisfy a critical analytical need as real-time methodologies become more widely used by oceanographers interested in small-scale temporal and spatial variability. It would also mean that a large amount of supplemental information would be available about the collected samples.

RADIONUCLIDES

There are three classes of radionuclides occurring in seawater:

- long-lived primary radionuclides and their daughters that have existed since their formation,
- cosmogenic radionuclides with relatively short half-lives that are formed continually by bombardment of matter with cosmic rays, and
- artificial radionuclides produced by human activities such as nuclear bombs and nuclear power stations.

The concentrations of these elements in seawater are extremely low. Radionuclides are primarily used as tracers to study water circulation, particle scavenging, exchange between water masses, and, in some cases, exchange between the ocean and the atmosphere or sediments.

Measurements of radionuclides in seawater have been used to study a variety of processes, including ocean mixing, cycling of materials, and carbon flux (by proxy). These measurements provide information on both process rates and mechanisms. Because of the unique and well-understood source functions of these elements, models of radionuclide behavior have often led to new understanding of the behavior of other chemically similar elements in the ocean.

Bomb-produced radionuclides in seawater have been used to determine ocean-mixing rates. The earliest measurements used cosmic ray-produced ^{14}C (radiocarbon) to establish that the time scale for mixing of the world ocean was on the order of 1000 years. Nuclear bomb testing from 1952 to 1962 injected large amounts of ^{14}C and ^{3}H (tritium) into the atmosphere. The atmospheric inventory for radiocarbon essentially doubled (Levin and Hesshaimer, 2000). Although these two radionuclides are both naturally occurring, bomb-testing elevated their surface ocean activities considerably above the natural background. Tracing transient bomb-produced ^{14}C and ^{3}H through the ocean has provided invaluable information concerning the rate of North Atlantic deep water movement and the rate of exchange of water through the main thermocline (Ostlund and Rooth, 1990; Broecker and Peng, 1982). Transient ^{14}C has also been used as a tracer of the transport for anthropogenic CO_2 into the ocean (Broecker et al., 1980) and will be valuable in validating ocean general circulation models that will be used to predict the fate of anthropogenic CO_2 in the ocean (Toggweiler et al., 1989; Guilderson et al., 2000).

Uranium—Thorium-Series Radionuclides

Naturally occurring radionuclides produced by the uranium and thorium decay series (Fig. 3.4) have been used for ocean mixing and biogeochemical studies. Sarmiento et al. (1990) used measurements of ^{228}Ra (half-life of 5.7 years) in the Atlantic to estimate the rate of nutrient input through the thermocline and the rate of carbon mineralization. Based on ^{228}Ra measurements in the equatorial Pacific, Ku et al. (1995) concluded that silicate limits biological production in this region. Moore et al. (1986) used ^{228}Ra to trace horizontal mixing and advection of water parcels that had passed through the Amazon mixing zone for over 1000 km into the Atlantic.

The short-lived radium isotopes have been used in coastal and nearshore studies. Moore (2000) used ^{223}Ra (half-life of 11 days) and ^{224}Ra (half-life of 3.7 days) to estimate rates of cross-shelf exchange in the South Atlantic Bight. Charette et al. (2001) used this pair to estimate the age of

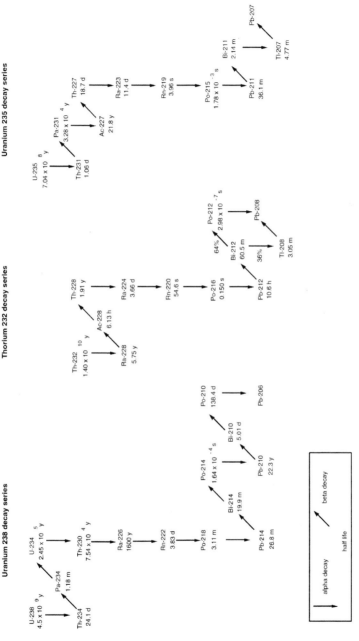

FIGURE 3.4. Isotopes of the uranium and thorium decay series.

water in a coastal pond. These two isotopes have great potential to solve a number of problems in estuarine and coastal oceanography.

Measurements of radionuclides are also used to determine removal mechanisms and controls for carbon and metal cycling in the ocean. For example, the removal of ^{234}Th from the euphotic zone is closely coupled to the vertical flux of particulate organic carbon. The deficiency of ^{234}Th with respect to its parent—^{238}U—in near-surface waters is used to estimate the export flux of particulate organic carbon (Buesseler, 1991). Measurements of ^{234}Th and ^{238}U in the upper water column provided the primary data relating to particulate carbon fluxes during JGOFS.

At present, there are no widely distributed certified reference materials containing all of the radionuclides in the uranium and thorium decay series. Such reference materials are needed to calibrate instruments that make radionuclide measurements and to compare analytical results from different laboratories. The most critical need is for reference materials in the ^{235}U decay series: ^{231}Pa, ^{227}Ac, and ^{223}Ra.

Because many of the radionuclide measurements in seawater require sample sizes of 20-200 liters, it is not practical to distribute true seawater reference materials containing these radionuclides. Furthermore, several of the radionuclide half-lives are only a few days to weeks. A different strategy is clearly required. The most reasonable approach is to prepare a reference material in the field by diluting solutions containing the long-lived parents of the short-lived radionuclides with a volume of seawater similar to the volume used for analyses.

Individual Standard Reference Materials containing ^{14}C, ^{3}H, and some naturally occurring uranium and thorium series radionuclides are available from the National Institute of Standards and Technology (NIST). These include:

- ^{238}U, ^{234}U, ^{235}U,
- ^{230}Th,
- ^{226}Ra, and ^{228}Ra.

No certified reference materials are currently available for:

- ^{231}Pa,
- ^{227}Ac,
- ^{223}Ra,
- ^{232}Th, and ^{228}Th,
- ^{224}Ra,
- ^{210}Pb, and
- ^{210}Po.

There are two approaches to providing the missing certified reference materials:

1. prepare additional individual solutions of ^{210}Pb, ^{210}Po, ^{231}Pa, ^{227}Ac, ^{232}Th, and ^{228}Th, or,
2. prepare three mixed solutions containing:
   (a) ^{238}U and ^{235}U with daughters in secular equilibrium through ^{226}Ra and ^{223}Ra,
   (b) ^{232}Th with daughters in secular equilibrium through ^{224}Ra, and
   (c) ^{210}Pb with daughters in secular equilibrium through ^{210}Po.

The presence of ^{222}Rn in the ^{238}U series makes the extension of this series through ^{210}Pb in the same solution very difficult due to the escape of radon gas. The advantage of the second (secular equilibrium) approach is that fewer solutions need to be prepared and the activities of the short half-life nuclides in these solutions will not change with time.

The committee thus recommends that NIST continue to provide reference materials for ^{14}C and ^{3}H as well as ^{238}U, ^{234}U, ^{235}U, ^{230}Th, ^{226}Ra, and ^{228}Ra. The committee further recommends the development of three new certified reference materials containing:

1. ^{238}U and ^{235}U with daughters in secular equilibrium through ^{226}Ra and ^{223}Ra,
2. ^{232}Th with daughters in secular equilibrium through ^{224}Ra, and
3. ^{210}Pb and ^{210}Po at secular equilibrium.

These acid solutions should contain on the order of 20 Bq/g of ^{238}U (1 Bq/g of ^{235}U), 20 Bq/g ^{232}Th, and 20 Bq/g ^{210}Pb.

CARBON ISOTOPES IN DISSOLVED INORGANIC CARBON (DIC)

The measurement of stable (^{12}C and ^{13}C) and radioactive (^{14}C) isotopes in DIC in the world's oceans provides information that can be used to study many aspects of the ocean carbon cycle. The distribution of both ^{13}C and ^{14}C in the ocean is governed by an interplay of biological and physical processes with carbonate chemistry. The normalization of ^{14}C to a constant ^{13}C as well as to a fixed point in time (1950) (Stuiver and Polach, 1977), removes biological and decay effects. This treatment allows ^{14}C to be used as a physical tracer while ^{13}C distributions can reveal information about biological processes. In surface waters, ^{13}C in DIC can be used to assess the uptake of anthropogenic CO_2 (Gruber and Keeling, 1999; Sonnerup et al., 1999; Quay et al., 2000). In deeper waters, ^{13}C can be used to study oxidation of organic carbon and its impact on nutrient concentrations (Lynch-Stieglitz et al., 1995; Mackensen et al., 1996). ^{14}C can be used to study the aging of ocean waters and to calculate pre-bomb surface water values (Toggweiler et al., 1989a; Key and Rubin, 2002). Although thousands of these measurements have been made recently on DIC as part of the WOCE and JGOFS Global Survey programs, and there are plans to repeat measurements over the next decades (NOAA, 2001), there are no formal seawater-based reference materials available for these isotopes.

One solution to the lack of standards for the carbon isotopes in DIC is to expand the use of the existing DIC concentration standard (Box 2.3) to include ^{13}C and possibly ^{14}C. This would be relatively simple: preliminary tests have shown the standards to provide reproducible ^{13}C measurements. Inclusion of ^{14}C in *any* reference material will require greater planning; while the reduction in sample size required for a radiocarbon measurement afforded by accelerator mass spectrometry (AMS) has greatly increased the scope of radiocarbon studies, it has also increased the likelihood of inadvertently contaminating a sample. Many oceanographic projects use radiocarbon as a tracer in experiments at sea and in the laboratory (e.g., for the measurement of oceanic productivity), and isolated spots or equipment can be accidentally contaminated. The radioactivities typically used in tracer experiments can be several million times modern levels and very small residual amounts can contaminate samples collected for the measurement of natural levels of ^{14}C. Table 3.2 details the concentration differences between natural levels and those typically found in productivity measurements. Because of the difficulties in methodology and the potential for cross-contamination, great care and plan-

TABLE 3.2 Abundance of ^{14}C in Natural Samples

Sample: Natural-level work	Activities (dpm/gC)
30,000 years old	0.3
20,000 years old	1.0
Modern	14
^{14}C productivity work	$0.6\text{-}6 \times 10^{12}$

NOTE: Activities are reported as disintegrations per minute/gram carbon (dpm/gC). The natural-level abundances are based on the activity of oxalic acid (NIST SRM 4990) and use the 5730 year half-life of ^{14}C (Karlen et al., 1964). The productivity work abundance is based on published JGOFS protocols for measuring primary productivity (Karl et al., 1996).

ning must be used when trying to prepare a seawater-based reference material for ^{14}C in DIC.

DISSOLVED ORGANIC MATTER (DOM)

Seawater contains on average about one mg/l of dissolved organic matter (DOM), which is by far the major form of organic material in the ocean. Seawater DOM contains a mass of carbon comparable to the total amount in atmospheric CO_2 and therefore is of interest as a major reservoir in the global carbon cycle. In addition, DOM is a key source of nutrition to marine organisms, with at least 50 percent of all marine primary production flowing through dissolved organic biochemicals to bacteria (the "microbial loop"). Marine DOM also complexes metals, attenuates light, promotes photochemical reactions and integrates ocean events on time scales ranging from seconds to millennia (Hansell and Carlson, 2002).

Studies of the amount, composition, and chemical reactivity of seawater DOM are challenging because of the low concentrations of the component molecules immersed in a background of 35,000 times as much dissolved salt. In addition, natural DOM mixtures are compositionally complex, with less than 10 percent of the component molecules identified to date as simple biochemicals. The only feasible method for quantifying total seawater DOM is to combust all of the component carbon to CO_2, which can be precisely measured as a proxy for total organic mass—seawater DOM contains approximately 50 weight percent of carbon. Although seawater DOM has been measured in this manner for almost 100 years, it was only in the early 1990s that the oceanographic community rigorously tested available methods for DOC analysis and came to a consensus concerning appropriate protocols and reference materials for routine analysis (Box 2.2). This effort (Sharp et al., 2002) is an outstanding example of how the oceanographic community can respond to an analytical challenge once the issues are clear and support is provided for workshops and (critical) environmentally pertinent reference materials.

There are two basic approaches used to characterize seawater DOM (Benner, 2002). The first of these is to directly analyze bulk compositions (e.g., elemental or isotopic compositions) or individual compounds in the sample without concentration. This approach requires high-sensitivity methods for either broad biochemical types (e.g., total amino acids or carbohydrates) or individual compounds, often by either spectroscopic or chromatographic methods coupled to electrochemical or mass spectrometric detectors. The latter type of molecular-level analyses are now feasible for measuring individual amino acids (Lindroth and Mopper, 1979), sugars (Skoog et al., 1999), and amino sugars (Kaiser and Benner,

2000) in less than 10 ml of seawater. The great information potential of organic compounds as source and reaction indicators, coupled with continued development of more sensitive and selective measurement methods based on mass spectrometric and electrochemical detection, suggests that direct molecular measurements will continue to grow in application.

The second basic approach to characterizing seawater DOM is to concentrate a fraction of the total mixture by chemical or physical means into either dry powder or concentrated solution. The solution can be analyzed using a wide array of methods (e.g., elemental analysis, biomarker analysis, mass, or nuclear magnetic resonance spectrometry) to which these isolates are amenable.

The first practical method for concentrating milligram amounts of seawater DOM was hydrophobic adsorption from acidified solutions onto a variety of substrates including charcoal and synthetic (e.g., polystyrene) resins and subsequent elution with a dilute base (e.g., 0.1 M sodium hydroxide [NaOH]) or methanol. Although the resulting "seawater humic substances" could be isolated as low-salt powders, this method involved severe pH conditions (pH = 1 to 13) and gave relatively low (5 to 25 percent of DOC) and chemically selective recoveries. Since the early 1990s (Benner et al., 1992), the DOM isolation method of choice has been tangential-flow ultrafiltration, which separates larger organic molecules from smaller sea salts and water. Although it still recovers less than half (25 to 40 percent) of total seawater DOM, ultrafiltration does not involve large pH fluctuations and is not intrinsically selective for different chemical compound types. A major drawback of ultrafiltration is that it is necessary to process 2000-5000 L of ocean water to recover 1g of isolate. At this scale, the isolation procedure requires special equipment, considerable operator skill, and approximately a week's time.

With the previous considerations in mind, the most practical approach to providing DOM reference materials for the oceanographic community may be to focus on providing small (1L) seawater samples for bulk chemical and molecular analysis. With some planning, it seems feasible that these reference materials could be the same as previously described for inorganic analysis. The previously discussed surface and deep seawater samples recommended for iron measurement might be particularly suitable for organic analysis. The preparation steps of filtration and acidification required for the iron samples have already been found to stabilize DOC concentrations for at least six months (Hedges et al., 1993). Although some component biochemicals may exhibit greater reactivity than total DOC in such samples, deep ocean samples contain molecules that have survived in situ for decades or more and should be relatively stable.

The stability of the organic components of surface seawater samples is more questionable, but can be tested over time for a wide variety of molecule classes.

Although iron samples must be stored in plastic (not glass) containers, some polymers such as Teflon® and high-density polyethylene are routinely used for storing DOC samples and should be suitable. Even if some plasticizer does leak from the containers, its components should not include (or interfere with the analysis of) commonly analyzed biochemicals such as amino acids and sugars. A major advantage of being able to analyze iron and DOM components of the same reference material is that the two measurements would complement each other, because the concentration and reactivity of iron is largely controlled by complexation with organic substances.

Although an available ultrafiltered DOM reference material would provide great benefit to oceanographers studying both trace organics and metals, the expense involved in obtaining sufficient material (kg) for broad distribution is disproportionately high versus that for more readily available water and sediment samples. A parallel example would be sinking particles from marine water columns, which are also of broad interest but difficult to obtain in sufficient amounts for broad distribution and use. Both ultrafiltered DOM and sinking marine particles might be candidates for a future reference material suite, or for workshop-based initiatives specific to these challenging sample types.

Existing Reference Materials for DOM

At present, soil derived humic matter and fulvic acids extracted from freshwater are available commercially and are commonly used to test techniques for DOM detection and also used as model compounds for trace metal chelation studies. The results obtained using these model compounds are frequently extrapolated to the natural environment and measurements on "real" samples provide evidence that this DOM is a good model compound. In the past, some investigators also made available organic matter isolated from marine environments using C18 resins. While these compounds come from aquatic sources, this isolation technique is chemically selective and isolates only a small percentage of oceanic DOM. Reference materials are not currently available for these compounds, which inhibits study of the role they play in a variety of oceanographic processes.

DISSOLVED GASES

Atmospheric gases can be divided into those that are relatively constant and those that vary in concentration. The most abundant (and thus relatively constant) components are nitrogen (as N_2), oxygen, and argon. The variable components are primarily water vapor and gases produced at least in part by human activities. These gases include CO_2, methane (CH_4), and carbon monoxide (CO). Both types of gases are exchanged between the atmosphere and the surface ocean with the result that they are generally at or near solubility equilibrium. They are then distributed throughout the ocean by down-welling water masses and large-scale ocean circulation. In ocean water, some gases behave conservatively, while others, such as O_2 and CO_2 are influenced by biological or other processes.

Dissolved gases in seawater have been used to examine the flux of greenhouse gases (e.g., CO_2, CH_4, nitrous oxide [N_2O]) across the air-sea interface as well as to trace and date water masses (e.g., chlorofluorocarbons [CFCs]). Biogenic gases like dimethylsulfide (DMS) are oxidized to sulfuric acid (H_2SO_4) in the atmosphere to produce cloud condensation nuclei. Gas-phase standards are generally available for most gases. No solution standards exist, however, for gases dissolved in seawater. This is largely due to the difficulty of preparing stable gas solution standards. At present, gas-phase standards are used to determine the reliability of measurement techniques, but these cannot be used to check the reliability of equilibration time and stripping techniques. There is a need for dissolved gas solution standards in order to put seawater measurements on a common basis. For non-greenhouse gases (noble gases) used to study ground water processes, the atmosphere can be used as a standard if reliable solubilities are available. Due to these obstacles, new gas-based reference materials are not recommended at this time.

4
Chemical Reference Materials for the Analysis of Particulate and Sediment Samples

RATIONALE FOR SEDIMENT AND PARTICULATE MATTER ANALYSES

Many of the analytes of interest for solid phase chemical reference materials are the same as those in seawater, but the need for and the preparation of reference materials for suspended particulate matter and sediments is quite different. The low concentrations of many seawater species and the presence of the salt matrix create particular difficulties for seawater analyses. However while sediments frequently have higher component concentrations than seawater, they also have more complicated matrices that may require unique analytical methods. A number of particulate inorganic and organic materials are employed as paleoceanographic proxies, tracers of terrestrial and marine input to the sea, measures of carbon export from the surface waters to the deep sea, and tracers of food-web processes. Some of the most important analytes are discussed below as they relate to important oceanographic research questions.

Major Bio-organic Elements

As discussed in Chapter 3, measurements of the major elements composing particulate organic matter are among the longest established, most fundamental, and broadly informative analyses in the marine sciences. This tradition is based on the fact that six elements (carbon, hydrogen, oxygen, nitrogen, sulfur, and phosphorus) make up essentially all the mass of marine organic matter and largely govern the processes by which these elements are cycled in the ocean. The rationale for the analysis of specific elements is presented below.

Carbon

Carbon is the most abundant element common to all organic substances (approximately 55 wt percent of average marine plankton) (Hedges et al., 2002) and has been used as the practical currency for measuring total organic matter concentrations in dissolved and particulate marine samples (MacKinnon, 1981). Carbon comprises two stable isotopes (^{12}C and ^{13}C), and one useful radioactive isotope (^{14}C), which together provide indicators of source, reaction history, age, and dynamics of the organic substances in which they occur (Raymond and Bauer, 2001). Carbon constitutes 27 wt percent of CO_2, a potent greenhouse gas and driver of climate change (Siegenthaler and Sarmiento, 1993; Sarmiento et al., 1998). At present, about 10 percent (3.5 Gt, 1 Gt = 10^{15} g) of all carbon actively cycling at the Earth's surface occurs in organic substances. The global rate of net photosynthesis (approximately 100 Gt C/yr) is sufficient to pass the entire mass of carbon currently in atmospheric CO_2 through living biomass in less than 10 years. Thus, any process or perturbation that affects the formation or remineralization rate of organic carbon can substantially influence the global carbon cycle and the potential for climate change within a human life span.

Carbon Isotopes

As mentioned above, the three isotopes of carbon are powerful tools for studying ocean processes, the carbon cycle and paleoceanography. ^{13}C and ^{14}C in biological materials and sediments are most commonly measured in bulk organic matter and in either bulk $CaCO_3$ or isolated foraminiferal shells. With the advent of isotope-ratio-monitoring gas chromatography mass spectrometry (irm-GCMS), sophisticated analytical isolation techniques, and accelerator mass spectrometry (AMS), carbon isotopes can now be measured in specific organic compounds isolated from sediments and particulate material.

In coastal areas, measurements of $\delta^{13}C$ in bulk organic matter can help identify the origins of organic material in sediments. In general, material produced using the dominant C_3 photosynthetic pathway has a value of $\delta^{13}C$ around –27 ‰ for terrestrial matter and around –20 ‰ for marine matter (Deines, 1980). The interpretation of such results is complicated because some plants use the C_4 photosynthetic pathway, which produces terrestrial matter with a $\delta^{13}C$ value near $-12‰$, and other plants use both pathways (CAM plants)[1] (Deines, 1980).

In the open ocean, $\delta^{13}C$ values of particulate organic matter in the surface ocean vary greatly. Isotopic variations have been linked to changes in temperature and latitude (Freeman and Hayes, 1992; Goericke and Fry, 1994). Fractionation is also a function of the concentration of CO_2 in the aqueous form, the $\delta^{13}C$ of the CO_2 (a function of temperature), and the fraction of the dissolved inorganic carbon pool taken up by an organism (Hayes, 2001; Laws et al., 2001). In addition to its use in studying organic matter, the measurement of $\delta^{13}C$ in the $CaCO_3$ of foraminifera is helping to evaluate the role of the deep ocean in global climate change. Long-term changes (i.e., over glacial cycles) in the mean ocean $\delta^{13}C$ and the isotopic relationship to nutrients over time can be similarly studied.

Radiocarbon measurements provide the accurate chronologies required to pinpoint causal events, either by determining sedimentation rates or by assigning an actual date to a specific event. AMS now provides the capability to measure radiocarbon routinely in small (0.5-1 mg) samples, making it possible to add chronologies to paleoceanographic studies of sediments as well as to use the natural ^{14}C cycle as a tracer of the transfer of carbon among the different global carbon pools. Measurement of radiocarbon in bulk organic matter establishes time histories within sediment columns and the rates of sediment accumulation in carbonate-poor areas of the ocean. The presence or absence of bomb-produced radiocarbon in bulk particulate and sedimentary matter provides clues as to how recently the material was actively involved in the carbon cycle.

Calibrations of $\Delta^{14}C$ in corals with U/Th series isotopes are extending the calendar chronology of ^{14}C back 20,000 to 40,000 years, allowing researchers to better study the events causing glacial/interglacial cycles (Bard et al., 1999). Radiocarbon measurements of bulk $CaCO_3$ can help provide information on the role of carbonate dissolution within sediments in the global carbon cycle (Broecker et al., 1980; Keir and Michel, 1993; Martin et al., 2000). Measurement of $\Delta^{14}C$ in coral bands provides a record of $\Delta^{14}C$ in the surface ocean (Druffel, 1995) and can be used to track the oceanic uptake of the atmospheric bomb pulse described in Chapter 3. Studies of the uptake of the bomb pulse provide useful constraints to ocean mixing models (Toggweiler et al., 1989a,b). The very detailed records of seasonal and annual fluctuations of δ ^{14}C in corals provide

---

[1]There are three major photosynthetic pathways: C_3, C_4 and CAM. Please see glossary for further detail. (Appendix C).

tighter constraints on ocean mixing in global carbon cycle models (Guilderson et al., 1998, 2000). Isotopes in specific organic compounds will be discussed separately.

Nitrogen, Phosphorus, and Sulfur

Nitrogen and phosphorus, in addition to being potentially limiting nutrients, provide an indication of the source and history of the organic materials in which they occur. Both elements are typically lost in preference to carbon during the digestion and bacterial degradation of plankton biomass, and therefore have potential use as indicators of substrate freshness. Because marine organic nitrogen occurs predominantly in protein, carbon to nitrogen (C:N) ratios can be indicative of the food quality of different organic mixtures. Since increased nitrogen richness is characteristic of marine mixtures, C:N ratios have also been used to discriminate plankton- from land-derived organic matter in coastal ocean zones. Such C:N-based assessment of organic source and food quality, however, must take into account that nitrogen can become elevated versus carbon in the course of biodegradation (Suess and Müller, 1980).

Neither nitrogen nor phosphorus has a long-lived naturally occurring radioisotope, and phosphorus has only one stable isotope. However, ^{15}N to ^{14}N ratios can be used under appropriate circumstances to detect nitrogen fixation, denitrification, cumulative uptake of upwelled nutrients, and passage of nitrogen up trophic levels (Goericke et al., 1994). Such stable isotope analyses have the potential to become a powerful tool in ecosystem studies that exploit the isotopic fractionation associated with the transfer of carbon and nitrogen between trophic levels (Gannes et al., 1997). Organic sulfur accounts for less than 1 wt percent of average marine plankton, is not a limiting nutrient in the ocean, and is difficult to quantify in marine samples due to interference by seawater sulfate.

Hydrogen and Oxygen

Hydrogen and oxygen, although major components of marine organic matter with multiple stable isotopes, have remained largely "orphan elements" with respect to biogeochemical studies. In large part, this analytical avoidance results from the challenge of measuring these elements in a sea of potentially interfering water in living cells, hygroscopic salts, and hydrous minerals. Current estimates of the elemental composition of marine plankton biomass based on measured biochemical compositions (Anderson, 1995) and spectroscopic analysis (Hedges et al., 2002) indicate a compositional formula near $C_{106}H_{177}O_{37}N_{17}S_{0.4}$. This study did not consider phosphorus (or its associated oxygen), as it undergoes no change in redox state during photosynthesis and remineralization and thus does not affect respiration demand. This revised formula corresponds to much lower hydrogen (8 wt percent) and oxygen (26 wt percent) contents than given in the Redfield-Ketchum-Richards equation (see Chapter 3) and requires approximately 10 percent more oxygen for complete respiration. Because the polysaccharides, proteins, and lipids that compose marine biomass have uniquely different elemental composition, their weight percent in plankton (and their remains) could be directly determined if the content of oxygen and hydrogen were measured in addition to carbon and nitrogen. The hydrogen content of organic matter is also of broad interest because it is positively related to the petroleum formation potential of marine sediments (Pederson et al., 1992; Gélinas et al., 2001a) and rocks (Demaison and Moore, 1980). Accurate measurements of the major elements composing marine organic matter offer a wealth of useful biogeochemical information, if appropriate methods can be developed and compared for widely available reference materials.

Specific Organic Compounds

The multitude of organic compounds present in the environment makes it impossible to discuss each compound, or even compound class. A general rationale for the matrix approach to preparation of reference materials is presented below, as well as a specific discussion of several classes of particular importance.

The last 50 years have seen a steadily increasing fraction of ocean measurements directed toward organic substances. This growing emphasis reflects the importance of organic molecules as the fundamental components of all living organisms and as key agents for the transfer and storage of the partially degraded remains of plants and animals in ocean waters and sediments. Organic molecules occur in a multitude of structures that are chemically unique and often persistent, and which carry a wealth of information about their biological sources and subsequent reaction histories. Almost all the organic matter in the ocean occurs as the incompletely degraded nonliving remains of marine organisms—living biomass is relatively minor. On average, organic substances account for about one in thirty carbons dissolved in seawater, four of five carbons sinking in particulate debris from the surface to the interior ocean, and one in five of all carbons preserved in marine sediments.

Lipid and Pigment Biomarkers

The synthesis, destruction, and vertical transport of organic matter in the marine environment all have important implications for the produc-

tivity of higher trophic levels (e.g., micronekton and fish) in coastal, pelagic, and benthic habitats. The major components of the marine planktonic food web include bacteria, *Archaea*, phytoplankton, and zooplankton. Knowledge of plankton community structure and its variability is
therefore essential for evaluating the predictive ability of complex food
web models and assessing ecological response to climate change (e.g., the
El Niño Southern Oscillation, the Pacific Decadal Oscillation, and the
North Atlantic Oscillation). Historically, the determination of the biomass and diversity of the microbial component (bacteria, *Archaea*, and
phytoplankton) within planktonic food webs posed a difficult problem
for oceanographers. A range of uncertainties, including the selectivity of
various culturing methods, overlapping size distributions, poor preservation properties, and the subjectivity of microscopic methods have encouraged the development of lipid biomarker approaches for characterizing
microbial communities.

Lipids are major constituents of all living cells and include a wide
range of functional and storage biomolecules, such as chlorophylls,
porphyrins, carotenoids, hopanoids, sterols, fatty alcohols, diterpenes,
ubiquinones, fatty acids, and waxes. This enormous structural diversity
reflects, in part, phylogenetic relationships, and consequently, specific
lipid compounds are frequently employed as biomarkers (Table 4.1).
Lipid biomarkers have varying degrees of taxonomic specificity. The
carotenoids and sterols of microalgae, for example, are typically used to
provide taxonomic distinction at the class level. Individual fatty acids,
hydrocarbons and sterols reflect their different plankton sources
(Wakeham and Lee, 1993) and help distinguish terrestrial from marine
origins. Bacteria add diagnostic branched-chain fatty acids to particles
they colonize (de Baar et al., 1983; Wakeham and Canuel, 1988).

Chlorophyll *a* (Chl *a*) functions as the primary light harvesting pigment in marine oxygenic phototrophs. Even though the C:Chl *a* ratio of
photoautotrophic cells varies considerably as a function of environmental
conditions and growth rate (Laws et al., 1983), measurements of Chl *a*
have been used extensively to estimate the biomass of photoautotrophic
microorganisms in the sea.

Cells are typically concentrated by filtration and extracted into an
organic solvent (usually acetone) after which, pigments are detected by
fluorescence or absorption spectroscopy, sometimes after chromatographic separation (Bidigare and Trees, 2000). The application of HPLC
to phytoplankton pigment analysis has lowered the uncertainty in the
measurement of Chl *a* and accessory carotenoids, since compounds are
physically separated and individually quantified.

Pigment distribution is useful for quantitative assessment of phytoplankton community composition, phytoplankton growth rate and

TABLE 4.1 Examples of Microbial Biomarkers and Potential Source Organisms (Volkman, 1986; Ourisson et al., 1987; Ratledge and Wilkinson, 1988; Mayer et al., 1989; Conte et al., 1994; Jeffrey et al., 1997; Béjà, et al., 2000: Kolber et al., 2000; Madigan et al., 2000)

Biomarker	Potential Source Organism(s)
Tetrapyrroles	
Divinyl chlorophylls a and b	*Prochlorococcus* spp.
Monovinyl chlorophyll b	Chlorophytes, prasinophytes
Chlorophylls c_1, c_2 and c_3	Chromophyte microalgae
Bacteriochlorophyll a	Anoxygenic photosynthetic bacteria
Carotenoids	
Peridinin	Dinoflagellates
Fucoxanthin	Diatoms
19'-butanoyloxyfucoxanthin	Pelagophytes
19'-hexanoyloxyfucoxanthin	Haptophytes
Alloxanthin	Cryptophytes
Prasinoxanthin	Prasinophytes
Lutein	Chlorophytes
Zeaxanthin	Cyanobacteria, chlorophytes
C_{20} isoprenoids	
Phytol	Photoautotrophs
All *trans*-retinal	*Proteobacteria*
Ether-linked lipids	*Archaea*
Sterols	
Dinosterol	Dinoflagellates
24-methylcholesta-5,22E-dien-3β-ol	Diatoms, Haptophytes
24-methylcholesta-5,24(28)-dien-3 β-ol	Diatoms
24-methyl cholest-5-en-3β-ol	Chlorophytes
C_{37-39} alkenones	*Emiliania huxleyi* and *Gephyrocapsa oceanica*
Hopanoids	Prokaryotes, including cyanobacteria
Diploptene, hopanoic acids	
Lipopolysaccharides (LPS)	Gram-negative bacteria
β-hydroxy-acids	
Polar lipid fatty acids	Bacteria, especially *Bacillus* spp.
Branched-chain C_{15} and C_{17} acids	
Peptidoglycan	Bacteria, mainly gram-positive strains
D-amino acids	

zooplankton grazing activity. Global maps of Chl *a* distribution are now available daily from satellite-based ocean color sensors such as SeaWiFS and MODIS. These high resolution (less than or equal to 1 km) Chl *a* images have provided new insight into the effect of physical and climate driven forcing on phytoplankton biomass and productivity at regional-to-global scales (Chavez et al., 1999; Behrenfeld et al., 2001; Seki et al., 2001).

Biomarker distributions can also provide important information on microbial metabolism. For example, the presence of Chl *a*, bacteriochlorophyll *a*, or *trans*-retinal in a culture or environmental sample would indicate microbial growth driven by oxygenic photosynthesis, anoxygenic photosynthesis, or photo-heterotrophy, respectively (Takaichi et al., 1990, 1991; Béjà et al., 2000; Kolber et al., 2000). Diagenetic products for many of the lipid biomarkers have been characterized structurally and used as "molecular fossils" to study ancient rocks, sediments, and petroleum (Lee and Wakeham, 1988; Wakeham and Lee, 1989; Peters and Moldowan, 1993; Kenig et al., 1994; Summons et al., 1999; Moldowan et al., 1996). These studies have demonstrated that biomarkers have utility for tracing the evolution of life in the geologic record, recreating the structure of ancient microbial communities, and determining which organisms contribute to hydrocarbon deposits.

The haptophyte microalga *Emiliania huxleyi* produces biomarkers in the form of long-chain (C_{37}, C_{38}, and C_{39}) alkenones (Brassell, 1993). Alkenones are well preserved in marine sediments and their molecular distributions and isotopic composition have been used to infer paleo-temperatures (Brassell, 1993) and pCO_2[2] values (Jasper et al., 1994), respectively. Unsaturation patterns in the alkenone series are related to the growth temperature of the haptophyte algae that produce these compounds (Brassell et al., 1986; Prahl and Wakeham, 1987), and hold great promise as indicators of absolute ocean paleotemperature.

A variety of molecular indicators of the freshness (and perhaps nutritional quality) of organic matter exists. For example, certain labile phytoplankton constituents, such as polyunsaturated fatty acids, are readily degraded in the environment or in herbivore guts, and are thus depleted in more degraded particles (de Baar et al., 1983; Wakeham and Canuel, 1988). Preferential loss of labile algal fatty acids resulting in the enrichment of more stable components in the products of heterotrophic metabolism has been observed in both field studies and laboratory feeding experiments (Prahl et al., 1985; Wakeham and Canuel, 1988; Harvey et al.,

---

[2]Partial pressure of CO_2. When partial pressure in one medium is higher thatn another there will be a net flow of gas from that phase to the other.

1987). A resistant organic matrix may protect otherwise labile lipids. For example, the resistant waxy coating common to land plants appears to physically shield terrestrial biomarkers, which are then selectively preserved, as opposed to their marine-derived counterparts in abyssal sediments (Gagosian et al., 1983; Wakeham et al., 1984; Volkman , 1986). The freshness of organic matter can also be influenced by an intimately associated mineral or organic matrix (Gordon and Millero, 1985; Hedges and Keil, 1995; Mayer, 1994).

Amino Acids and Sugars

Amino acids are structural components of proteins and compose the largest reservoir of organic nitrogen in most organisms. They make up a major fraction of characterized carbon in marine particulate matter and are useful indicators of decomposition and transport in the marine environment. Their natural occurrence in marine plankton (Degens and Mopper, 1976; Tanoue et al., 1982; Cowie and Hedges, 1996), suspended and sinking particles (Lee and Cronin, 1982, 1984; Ittekkot et al., 1984a, b; Cowie and Hedges, 1992; Cowie et al., 1992), and sediments (Henrichs and Farrington, 1984; Henrichs et al., 1984; Dauwe and Middleburg, 1998) is frequently used to elucidate the biogeochemical behavior of marine organic matter, particularly the diagenetic state. Chromatographically measurable carbohydrates constitute a maximum of 20 to 40 percent of plankton carbon, 13 percent of sinking particulate organic carbon (POC), 15 percent of suspended POC, and 13 percent of sedimentary carbon (Hernes et al., 1996). Amino acids and carbohydrates can be used to elucidate the fate of organic carbon and to infer past marine sources and oceanic conditions.

Although generally not as species-specific as lipids and pigments, compositional differences among carbohydrates and amino acids can also reflect biological sources. For example, some siliceous diatoms are characterized by elevated concentrations of glycine and fucose while organisms with carbonate tests often exhibit unusually high concentrations of aspartic acid and arabinose (Hecky et al., 1973; Carter and Mitterer, 1978; Ittekkot et al., 1984a,b). Although both plankton and bacteria are characterized by high relative abundances of ribose and fucose (Cowie and Hedges, 1984), O-methyl sugars and uronic acids may provide a means of discriminating between these two biological sources (Mopper and Larsson, 1978; Bergamaschi et al., 1999). In addition, muramic acid can serve as a specific marker for bacteria (Lee et al., 1983).

High relative concentrations of nitrogen and carbon in the form of amino acids (Whelan, 1977; Lee and Cronin, 1984; Cowie and Hedges, 1992) and carbohydrates (Cowie et al., 1992; Cowie and Hedges, 1992,

1984) indicate relatively undegraded organic matter. Conversely, high relative concentrations of characteristic diagenetic products, such as the non-protein amino acids ornithine and β-alanine can indicate the presence of bacterially degraded material (Lee and Cronin, 1982, 1984; Ittekkot et al., 1984a,b). The combined use of multiple amino acid- and carbohydrate-based diagenetic indicators provides a sensitive and consistent means of comparing the state of organic matter in aquatic environments (Hedges et al., 1999). Preservation of certain amino acids and sugars occurs in both siliceous and carbonate tests. These can be sensitively detected in part because amino acid composition of fresh materials is so uniform.

Carbon and Nitrogen Isotopes in Specific Compounds

Although carbon isotopic measurements of bulk organic material can be very informative, much more information can be obtained from the isotopic characterization of specific compound classes or individual organic compounds isolated from bulk material. Such compound-specific isotope analyses (CSIA) can target individual molecules known to be associated primarily with specific processes (e.g., photosynthesis) or with specific sources.

The stable isotopic compositions of algal organic matter can provide important insight into the environmental conditions under which carbon and nitrogen fixation occur (Hayes et al., 1990; Goericke et al., 1994). Determination of $\delta^{13}C$ and $\delta^{15}N$ provides valuable tools for assessing the rates of ancient biological processes such as phytoplankton growth and NO_3 uptake (Altabet and Francois, 1994; Bidigare et al., 1999). Application of stable isotopes to the modern ocean and sedimentary record, however, is confounded by the presence of non-algal carbon and nitrogen biomass. To resolve this problem, CSIA has been employed as a means of estimating the $\delta^{13}C$ and $\delta^{15}N$ of modern and ancient phytoplankton populations (Jasper et al., 1994; Bidigare et al., 1997, 1999; Sachs and Repeta, 1999).

Researchers are now investigating the potential for dating targeted compound classes and individual compounds to understand the true chronology of sediments. Pearson et al. (2000, 2001) measured the radiocarbon content of individual sterols and other lipids isolated from sediments. While most of the compounds were derived from marine photosynthesis or subsequent remineralization, data from the archeal isoprenoids were consistent with chemoautotrophic growth below the euphotic zone. Eglinton et al. (1997) found significant variation in the age of individual compounds isolated from the same sediment, indicating that older components either have a different inorganic carbon source or are stored for appreciably longer time periods in "upstream" reservoirs.

Radionuclides

Measurements of radionuclides and metals in marine sediments and particulate matter are conducted for a variety of purposes, including the determination of sedimentation rates, trace metal and radionuclide fluxes through the water column, enrichment of metals in specific phases of the sediments, and examination of new sedimentary phases produced after sediment deposition. Such studies address fundamental questions concerning the chronology of deep-sea and near-shore sedimentary deposits, removal mechanisms and cycling of metals in the ocean, and diagenesis within deep-sea sediments.

Numerous radionuclides have been applied to marine sedimentary problems. These are generally grouped into cosmic-ray produced (cosmogenic) nuclides (^{14}C, ^{10}Be, ^{7}Be, ^{26}Al), nuclear bomb-produced (fallout) nuclides (^{137}Cs, plutonium isotopes, ^{241}Am), and naturally occurring nuclides ultimately derived from the decay of ^{238}U, ^{235}U, and ^{232}Th parents.

^{230}Th and ^{231}Pa are ubiquitous components of recently deposited deep-sea sediments because they are produced uniformly throughout the ocean from the decay of dissolved uranium isotopes and they are actively collected onto sinking particles. The distribution with depth of these nuclides in deep-sea sediments may be modeled to estimate rates of sedimentation extending over the past 200 to 300 thousand years. These techniques complement ^{14}C dating methods that only extend to about 40 thousand years before the present.

Other techniques are used for shorter time scales, including the measurement of the ^{226}Ra:Ba ratio in barite extracted from sediments (Paytan et al., 1996). This technique has a time scale of about five thousand years. Alternately, assessments of rapid sedimentation and bioturbation on time scales of days to centuries require shorter half-life nuclides such as ^{210}Pb, ^{228}Th, ^{234}Th, and ^{222}Rn.

INFLUENCE OF MATRIX COMPOSITION
ON CHEMICAL DETERMINATIONS

Several major matrix types are found in marine particles and sediments. Marine organisms surround themselves with tough polymeric organic cell walls and/or with opal or calcium carbonate tests. These contrasting matrices respond differently to various analytical methods. In sediments, the remains of these organisms combine with clay minerals to form a heterogeneous mixture. In this section, the influence of these matrices on analyte quantification are discussed.

Major Bio-organic Elements

The organic carbon, hydrogen, and nitrogen content of marine samples is usually measured by quantitative combustion of a weighed amount of solid (or liquid) material into volatile gases whose concentration can then be measured in a CHN analyzer, equipped with a nonspecific thermal-conductivity detector. Typically, carbon is measured as CO_2, nitrogen as N_2 (after reduction of nitrogen oxides), and hydrogen as H_2O. In the simple case of dry, purely organic samples, this method is fast, accurate, precise, and requires only 1 mg or less of sample. Unfortunately, dry, purely organic materials are almost never obtained in marine samples because many common types of phytoplankton and zooplankton secrete calcium carbonate (e.g., coccoliths and foraminifera) or opal (e.g., diatoms and radiolarians) tests. Even samples of living plankton taken directly from surface ocean waters characteristically contain 20 to 60 weight percent of mineral matter (Parsons et al., 1977). Due to the relative lability of the organic versus the mineral components of plankton remains, the weight percent of mineral matter in marine particles increases with depth and degradation, reaching typical values near 90 wt percent in particles raining to the ocean floor and more than 99 wt percent for those particulate materials that make up marine sediments (e.g., Wakeham et al., 1997).

The minerals that make up the increasingly predominant inorganic fraction include:

- hydrous aluminosilicate clays of largely terrigenous origin usually predominating along continental margins and in open-ocean red clay deposits,
- calcite and aragonite remains that are most abundant in carbonate oozes accumulating on ocean topographic highs such as mid-ocean ridges, and
- opal oozes formed largely from diatom debris accumulating beneath highly productive surface ocean waters in upwelling zones along continental margins and at equatorial and high latitudes.

Because samples of suspended, sinking, and sedimentary marine particles often contain appreciable amounts of clay minerals, carbonate, and opal, any quantification method for organic elements must avoid interference from these mineral types.

Carbon

Organic carbon is by far the most commonly quantified organic element in marine samples, for which $CaCO_3$ is the most typically encountered interference. This problem occurs because carbonates decompose upon heating, releasing CO_2, the same gas that CHN analyzers measure after the oxidation of organic matter. Early attempts to distinguish these two sources by selectively breaking down $CaCO_3$ at a lower temperature (400 to 650 °C) than is used to oxidize organic matter (greater than 1000 °C) failed because the temperature ranges over which CO_2 is generated by these two degradation processes overlap substantially (Froelich, 1980; Yamamuro and Kayanne, 1995). The only practical approach to analyzing organic and inorganic carbon separately has been to acidify samples at low temperature so that inorganic carbon can be quantitatively evolved as CO_2 while organic carbon remains behind relatively unaltered in the solid sample. For example, Weliky et al. (1983) directly quantified inorganic carbon by adding hydrochloric acid (HCl) to a known mass of sediment in a closed container and measuring the released CO_2. Although organic carbon can then be determined by the difference between total carbon (measured by combusting an untreated sample at high temperature) and inorganic carbon, this difference approach is imprecise for small amounts of organic carbon in carbonate-rich samples.

Most methods for directly measuring POC involve the use of a non-oxidizing acid (e.g., HCl, H_3PO_4, H_2SO_4) to remove carbonate prior to quantifying the organic carbon using high temperature combustion. Early methods of this type often involved preparative treatment of solid samples in large volumes of acid solution, after which the remaining solids were separated (by filtration or centrifugation), rinsed, dried, and weighed for high temperature combustion. Both organic matter and non-carbonate mineral matter could be dissolved and lost with the discarded treatment water (Froelich, 1980; Hedges and Stern, 1984), however. Such losses occur to varying extents depending on sample type and are especially pronounced for carbonate oozes (Yamamuro and Kayanne, 1995).

Once this problem became clear, later methods for organic carbon analysis used procedures in which only volatile substances were allowed to leave the sample, and any treatment water was retained. Acidification was accomplished using either HCl vapor (Hedges and Stern, 1984) or liquid acids added to watertight containers (Verado et al., 1990; Nieuwenhuize et al., 1994). Even in the absence of appreciable carbonate, open ocean clays often contain such low concentrations of organic carbon (approximately 0.1 to 0.3 wt percent) that precise measurement is difficult because of the small carbon blank intrinsic to all CHN measurements (Nieuwenhuize et al., 1994). In theory, organic and inorganic carbon can now be measured accurately for all major sample matrix types. Whether this is in fact the case across the oceanographic community is unknown because no intercomparisons have been reported for different mineral matrix types.

Carbon Isotopes

Many of the same matrix effects mentioned above affect stable and radiocarbon isotopes. An additional constraint is imposed on the analysis of organic carbon for $\delta^{13}C$ measurements because the CO_2 generated during combustion must be pure enough to introduce to an isotope ratio mass spectrometer. This has not presented a problem in practice because the small amount required for stable isotope analysis permits the isolation of very clean CO_2. The size of the sample required for AMS analysis (40-80 μmol C) for radiocarbon measurements presents an even greater analytical challenge. With some methods (e.g., H_2SO_3 dissolution), it is very difficult to determine the point at which all of the carbonate has been removed from a large sample (Verado et al., 1990). Other methods leave residual salts that produce large amounts of undesirable gases upon sample combustion. In addition, many of the salts are hygroscopic and the decarbonated samples must be cleaned prior to combustion. In practice, many laboratories resort to acidification, followed by rinsing. The resulting removal of some of the organic matter may affect the radiocarbon age measured on sediments.

Nitrogen

Nitrogen presents its own unique analytical challenges in solid marine samples. First and foremost, all forms of inorganic and organic nitrogen are converted in CHN analyzers to N_2 gas, which thus represents total, rather than strictly organic nitrogen. Although this is not a major problem for most organic-rich samples, open ocean clays can have concentrations of organic carbon substantially less than 1 weight percent (Suess and Müller, 1980). Thus with such low fractions of organic matter, a major fraction of the total nitrogen in these clays is derived from fixed ammonia rather than organic nitrogen. Acid treatments, used to remove inorganic carbon prior to CHN analysis, have a variable effect on measured nitrogen contents, further complicating analysis. A second complication is the variable effect on measured nitrogen content that is caused by the acid treatment frequently used to remove inorganic carbon prior to CHN analysis. Although this complication is not always observed (Hedges and Stern, 1984), comparisons of total nitrogen content across different sediment types are preferably done on untreated samples (Gélinas et al., 2001b).

Hydrogen

CHN analysis of most marine particle samples has typically ignored hydrogen because at temperatures over 1000°C generated in CHN analyzers, both opal ($SiO_2 \cdot nH_2O$) and hydrous aluminosilicates decompose with the evolution of water. For most marine sediments, resulting yields of mineral-derived water greatly exceed those generated by organic matter oxidation, making organic hydrogen analyses impossible. Acidification of $CaCO_3$ with HCl yields $CaCl_2 \cdot nH_2O$, an extremely hygroscopic salt that releases copious amounts of water upon heating, further complicating analysis. Finally, many types of organic matter hold water tenaciously, which elevates organic hydrogen values unless the sample is heated to high temperatures, which can cause biochemical decomposition. These complications in directly measuring organic hydrogen in marine materials is unfortunate given the importance of this element as an indicator of biochemical composition, biological oxygen demand, and petroleum-forming potential.

Oxygen

Oxygen lies somewhere between organic carbon and hydrogen with respect to its analytical difficulty. In samples that are largely organic (e.g., marine macrophytes and mineral-poor plankton), organic oxygen content can be estimated by subtracting the mass of carbon, hydrogen, nitrogen, and "ash" from the total sample mass. This mass difference approach, however, assumes the other three organic elements can be quantified accurately which is often in question (as detailed in previous discussion). In addition, "ash" is operationally defined as the mass of mineral matter that remains *after* all the organic component has been oxidized away by holding the sample under air in a furnace at a high temperature (400 to 600 °C). Residual ash is assumed to have the same mass as the *original* inorganic matter in the sample, so that this value can be subtracted from the initial total sample mass to estimate the initial organic mass. However, severe heating during ashing can cause inorganic components of the sample to either lose (e.g., by evolving CO_2, SO_2, and H_2O) or gain (e.g., by adding oxygen to iron and sulfur) mass, making accurate ash corrections difficult for mineral-rich samples. An alternate, and potentially much more accurate approach, is to measure organic oxygen evolved pyrolytically as CO at a temperature near 1000 °C. This is typically done in an oxygen-free gas stream within a modified CHN analyzer, where the evolved oxygen compounds are converted over an activated carbon bed to CO, which is then measured with a thermal conductivity detector. In practice, this direct oxygen method is seldom used. When it is, precision is typically poorer than for carbon, hydrogen, and nitrogen (Cheng-Tung et al., 1996).

Specific Organic Compounds

Overview of Analytical Techniques

Measuring the type and amount of organic substances in marine systems involves major challenges, some of which are unique to large, covalently-bonded molecules. Primary among these analytical hurdles is that essentially all marine biomass, seawater, or sediment samples contain thousands of different molecules that cannot all be analyzed together. Even the most concerted efforts to individually quantify the biochemicals in marine samples have accounted for only 80 percent of the total organic matter in living plankton, and less than 25 percent of the molecules in seawater and sediments (Wakeham et al., 1997). Such inventories remain incomplete in large part because each biochemical class (e.g., amino acids, sugars, lipids, and pigments) requires a different analytical procedure. As a result, a comprehensive survey across all the quantitatively important molecular types is logistically impractical.

The challenge of specific molecular analysis is compounded by the constraint that structurally definitive measurements of individual organic compounds are still primarily limited to relatively small (less than 1000 atomic mass units [amu]) organic molecules that can be chromatographically separated and quantified against commercially available chemical standards. In contrast to lipids, most organic substances in ocean samples exist in large molecules (e.g., proteins and polysaccharides) that must be chemically broken down into small structural units (e.g., amino acids and sugars) in order to be extracted from the sample matrix, identified, and quantified. Even in the simple case of a pure macromolecule, chemical treatments (e.g., hydrolysis or oxidation) severe enough to efficiently release small structural units often alter these desired products before the parent macromolecule can be completely broken down. Quantifying multiple compound types in organic mixtures is even more difficult because the reaction conditions for maximum conversion to measurable structural units seldom overlap, and the freed structural units can react with each other. Hydrolysis-resistant macrobiomolecules such as lignins, algaenans, and plant chars require extremely severe breakdown reactions such as pyrolysis and nitric acid oxidation, which typically yield only a fraction of the parent material in greatly altered form.

Finally, but critically, the matrix of a sample can have a profound effect on measurements of individual organic molecules within it. Such matrix effects can take many forms and be complexly interrelated. Organic substances in seawater and sinking or sedimentary particles are intimately associated with the much greater amounts of inorganic materials such as sea salt and mineral grains present in the marine environment. It is necessary, therefore, that the organic moiety either be isolated before analysis, or that it be broken down within (and extracted from) these matrices. Unfortunately, physical or chemical separation of organic substances from inorganic matrices of marine samples is almost never complete and often involves combinations of severe reagents (e.g., HF:HCl or NaOH:HCl) that can lead to appreciable chemical alteration. Thus, many types of useful characterizations (e.g., infrared, Raman, or NMR spectroscopy) that are possible for purely organic samples are infeasible for the same dilute components of natural mixtures. Matrix effects that can occur when the target analyte is generated within and extracted from an intact sample include:

- reaction of added chemical agents with the matrix instead of with the desired analyte precursor,
- physical shielding of the analyte or its precursor from the reagent or extraction solvent,
- generation of chemicals from the sample that react with the target analyte,
- uptake of the released analyte by the matrix, and
- coextraction of a matrix component that subsequently interferes with analysis of the target material.

Analysis of specific compounds in marine samples is currently constrained to a restricted number of organic components within a complicated mixture of organic matter of widely varying composition. This restriction is similar in many ways to the problem of determining the "speciation" (exact chemical form) of inorganic elements. Although it may be relatively straightforward to quantify the total amount of a component element (e.g., organic carbon) in a sample, it is extremely challenging to define its actual chemical forms and proximate environments as they existed at the time the sample was taken. However, speciation of an element or molecule is often the main determinant of both its reactivity and information potential. Even though molecular-level organic analyses are incomplete and capture only a fraction of the total information available in the intact sample, they nevertheless can provide unique and invaluable insights in both the form of qualitative (e.g., ratios and homologue fingerprints) and of quantitative parameters (e.g., amounts, ages, and fluxes).

Given this wealth of potential information, the many types of molecules that can be analyzed, and the relatively short time (approximately 50 years) that a small number of biogeochemists have studied the ocean, it is understandable that a diverse array of analytical methods have evolved around individual types of organic compounds. The current state of the art in marine organic geochemistry is that each lab group typically uses its own home-grown method for a given compound class. Often, but not always, the procedure is traceable to a rigorous methods paper in which analyte identities have been confirmed and their recoveries from a limited set of sample matrices have been systematically maximized. Seldom, however, is the analytical method rigorously tested for all the matrix types to which it is eventually applied, nor are the consequences of this adaptation of a published procedure always studied. Because of the many individual complex steps by which the target molecules are generated, extracted, isolated, derivatized, and eventually quantified, the overall outcome of an organic analysis can be very sensitive to small differences in laboratory techniques and available equipment. Studies of analytical drift over time within individual labs, as well as rigorous intercomparisons among different labs using contrasting analytical procedures for the same compound types are rare. The overall result of this analytical discontinuity is that it can be extremely difficult to bring data from multiple sources together in the study of extended environmental and temporal trends that are increasingly critical at this time of unprecedented global change. Analytical results from different organic measurements, however, could be effectively related by routine comparisons to reference materials that represent the matrices encountered in the environmental samples being analyzed (e.g., Fig. 2.1).

Pigment and Lipid Biomarkers

The analysis of pigment biomarkers in microbial cultures and marine particles is relatively straightforward, with the exception of pigments in armored dinoflagellates, heavily silicified diatoms and heavily walled green algae, which can be notoriously difficult to extract (Wright et al., 1997). The extraction of lipid biomarkers from certain sediment matrices is even more problematic. This is especially true for carbonate sediments, where extraction efficiencies and molecular distributions can vary significantly and can depend upon the type of extraction techniques used (e.g., wet or dry solvent extraction and acid or base hydrolysis, followed by solvent extraction).

Amino Acids

Amino acids are generally considered to be labile and easily analyzed in most marine matrices. However, recent studies have shown that a significant percentage of the proteinaceous component of marine particulate matter is not accessible using traditional methods (Hedges et al., 2001). The actual source of the uncharacterized fraction is not known, but at least part of it is undoubtedly associated with mineral matrices. For example, the acidic amino acids—aspartic and glutamic acid—are enriched in carbonate skeletons (Degens and Mapper, 1976) and are preferentially adsorbed onto carbonates (Carter and Mitterer, 1978). Organic molecules serve as a template for the secretion of all biogenic minerals (Lowenstam and Weiner, 1989). Usually rich in glycoprotein, this organic matter can be occluded in the skeletal matrix, which protects it from degradation unless the mineral is dissolved (King, 1974; Robbins and Brew, 1990). Sediments with high biomineral concentrations may contain a large fraction of organic carbon in skeletal matrices (Ingalls et al., in press). Organic matter preserved in biominerals currently serves as a proxy for numerous paleoenvironmental studies. Some examples are: the calculation of historic surface water pCO_2 (Maslin et al., 1996), estimates of $\delta^{15}N$-based paleoproductivity (Sigman et al., 1999) and species identification (King, 1974; Robbins and Brew, 1990; Collins et al., 1991; Endo et al., 1995).

Carbohydrates

Molecular-level carbohydrate analyses are challenging for several reasons. First, acid hydrolysis releases simple sugars which are prone to dehydrate (or otherwise degrade) before the parent polysaccharide can be completely depolymerized. In addition, hydrolysis and isolation conditions for neutral carbohydrates are often inappropriate for acidic and basic sugars (Benner, 2002), which may also interact differently with matrix minerals due to their charges. Finally, sugars also can condense with amines, including mineral-derived ammonia, to form highly altered nitrogenous products that escape measurement. A recent study of mineral-rich plankton tow materials (Hedges et al., 2001) indicated that only 25 percent of carbohydrate-like material (as determined by ^{13}C NMR) was measurable chromatographically as neutral sugars.

Radioisotopes

Measurement of radionuclides in marine solids may be accomplished by either destructive or nondestructive techniques. The nondestructive technique involves direct measurement of gamma rays emitted during radionuclide decay. The interaction of these gamma rays with the sample is a function of the energy of the gamma-ray and the composition of the sample; low energy gamma rays are strongly absorbed by sedimentary minerals. For example, ^{210}Pb is often measured by direct counting of low energy gamma rays (46.5 keV) that are partially absorbed by minerals within the sample matrix. The fraction of the gamma rays absorbed by the sample is a function of the sample matrix. There are methods to account for self-absorption, but there are no widely distributed reference materials to test these corrections. Similar problems exist for measuring gamma rays from other radionuclides.

Measurement of radionuclides by destructive analyses can involve combustion, fusion, leaching, or acid-dissolution of the samples. The choice of technique depends on both the radionuclide of interest and the sample matrix. Techniques developed for one matrix may not work for other matrices. In addition, many radionuclides in sediments exist both as surface-bound components adsorbed by particles sinking through the water column, and also as a structural component within physically intact mineral grains. These detrital structural components generate a background of supported activity that does not change with time. Thus, to use nuclides such as ^{230}Th and ^{231}Pa for dating, the radioactivity derived from uranium decay in the solid (supported activity) must be subtracted from that derived from adsorbed ^{230}Th and ^{231}Pa (unsupported activity). Different analytical techniques for these physically dissimilar phases may lead to different corrections for supported activity, which has yet to be separated with convincing conclusions.

For the ^{226}Ra:Ba ratio in barite technique, the results depend on the isolation of a pure barite sample and the assumption that the separation techniques do not alter the ^{226}Ra:Ba ratio of the barite. Applying different techniques to reference sediments could strengthen this technique.

REFERENCE MATERIALS CURRENTLY AVAILABLE FOR THE ANALYSIS OF SEDIMENT AND PARTICULATE SAMPLES

Solid Matrix Reference Materials

A selected list of reference materials (sediments as well as biological tissues) distributed by several Canadian, U.S., and E.U. sources shows a wide range of solid samples that could be used for comparative analysis of major organic elements (Table 4.2). These materials are widely available and have been analyzed for at least some constituents. In addition, these materials are homogeneous and can be expected to exhibit stable compositions over time. All of the thirty or so listed reference materials, however, lack precise measurements of the major bio-organic element compositions of the materials. The trace organic components that are certified, such as PAH, PCBs, pesticides, and hydrocarbons, do not constrain the concentrations of the organic elements. These samples do have potential for use as consensus reference materials, but only after elemental data are provided by different labs for selected samples using contrasting and well-defined analytical procedures.

The matrices and sources of the sediments listed in Table 4.2 are sometimes unclear. Those that are known are highly weighted toward clastic (quartz- and aluminosilicate-rich) marine sediments from coastal environments. Some of these reference materials, such as MESS-3 (NRC-Canada), MAG-1 (USGS) and the Arabian Sea and Pacific Ocean samples (IAEA 315, and 368), could provide excellent examples of clastic marine sediment representing the main repositories of organic matter in the ocean (Hedges and Keil, 1995). The listed materials fail to include both open-ocean opal and carbonate oozes, as well as pelagic red clays.

The biological reference materials listed in Table 4.2 represent a small subset of all the biomass samples available for analysis, though it is representative of the emphasis placed on animals versus macrophytes. Reference materials for plankton are particularly limited, although samples of the freshwater green alga, *Chlorella*, are available for trace element analysis from the Slovak Institute of Radioecology and Japan's National Institute of Environmental Studies. Notably, no reference materials appear to be available for marine phytoplankton of any type. This dearth of marine biological material for elemental analysis is unfortunate because it fails to represent the opal and carbonate matrices that can greatly complicate carbon, oxygen, and hydrogen analyses.

Carbon Isotopes

Organic and carbonate reference materials for $\delta^{13}C$ and $\Delta^{14}C$ are both available and widely used for isotopic analysis. The stable carbon isotope reference materials currently available are useful for instrument calibration and limited methods verification. These materials are available as powders for carbonates and in a wide range of forms for organic materials (e.g., oil, graphite, etc.) (Table 4.3). There is one certified primary standard for radiocarbon (NIST SRM 4990C) and a number of additional reference materials are available from the International Atomic Energy Agency (IAEA) in Vienna, and from Marian Scott at the University of Glasgow (Table 4.4, Appendix E). The FIRI samples represent a study in progress by an international group of radiocarbon researchers (Bryant et al., 2001). Samples from these sources are available primarily as discrete organic or carbonate materials and do not represent materials (such as

TABLE 4.2 Selected Solid Reference Materials Currently Available from Canadian, U.S., and European Sources with Potential Utility for Marine Organic Studies (See Appendix E for information about obtaining these materials.)

Identifier	Type	Matrix
Selected materials distributed by the National Research Council of Canada (NRC)		
MESS-3	Sed	clastic marine sediment
PACS-2	Sed	clastic marine sediment
HISS-1	Sed	clastic marine sediment
HS-3B	Sed	clastic marine sediment
HS-4B	Sed	clastic marine sediment
HS-5	Sed	clastic marine sediment
HS-6	Sed	clastic marine sediment
SES-1	Sed	clastic estuarine sediment
CARP-1	Bio	whole carp residue
DOLT-1	Bio	dogfish muscle
DORM-2	Bio	dogfish muscle
LUTS-1	Bio	lobster hepatopancreas
TORT-2	Bio	defatted lobster pancreas
Selected materials distributed by the U.S. National Institute of Standards and Technology		
SRM 1648	Dust	urban particulate matter
SRM 1649A	Dust	urban dust
SRM 1650A	Soot	diesel particulate matter
SRM 1939	Sed	clastic river sediment
SRM 1941B	Sed	clastic marine sediment
SRM 1944	Sed	clastic marine sediment
SRM 1945	Bio	whale blubber
Selected materials distributed by the International Atomic Energy Agency (IAEA), Austria		
IAEA-384	Sed	marine sediment
IAEA-315	Sed	clastic marine sediment
IAEA-368	Sed	marine sediment
IAEA-356	Sed	clastic marine sediment
IAEA-383	Sed	clastic marine sediment
IAEA-408	Sed	clastic marine sediment
Selected material distributed by the U.S. Geological Survey (USGS)		
MAG-1	Sed	clastic marine sediment

Abbreviations: %OC = weight percent organic carbon, Sed = sediment, Bio = biological, E = elemental, TE = trace element, PAH = polynuclear aromatic hydrocarbons, PP = PAH + PCB, PPP = PAH + PCB + pesticides, H = hydrocarbons, S = sterols, and ND = not determined, U = Unknown.

Source	Form	Analyte	%OC
Beaufort Sea	dry powder	TE	~2
Esquimalt BC harbor	dry powder	TE	~3.3
Hibernia shelf, NFL	dry powder	TE	ND
Nova Scotia harbor	dry powder	PAH	ND
Nova Scotia harbor	dry powder	PAH	ND
Nova Scotia harbor	dry powder	PAH	ND
Nova Scotia harbor	dry powder	PAH	ND
U	dry powder	PAH	ND
Lake Huron	water slurry	PCB	ND
U	dry powder	TE	ND
U	dry powder	TE	ND
U	water slurry	TE	ND
U	dry powder	TE	ND
St. Louis atmosphere	dry powder	PAH	ND
Washington DC air	dry powder	PAH	ND
heavy duty engine	dry powder	PAH	ND
U	dry powder	PCB	ND
U	dry powder	PP	ND
U	dry powder	PPP	ND
U	frozen tissue	PP	ND
Fangatufa Atoll	dry powder	RN	ND
Arabian Sea	dry powder	RN	ND
Pacific Ocean	dry powder	RN	ND
U	dry powder	TE	ND
U	dry powder	PHS	ND
Targus Estuary	dry powder	PHS	ND
Gulf of Mexico	dry powder	E	2.15

TABLE 4.3 Carbon Isotopic Composition of Selected Carbon–bearing Isotopic Reference Materials (Coplen et al., 2001)

Identifier	Substance	$\delta^{13}C$	Reference
NBS 18	CaCO$_3$ (carbonate)	−5.01 ± 0.06	Stichler, 1995; Coplen et al., 1983
NBS 19	CaCO$_3$ (calcite)	+1.95	Hut, 1987
IAEA–CO–1	CaCO$_3$ (marble)	+2.48 ± 0.03	Stichler, 1995
IAEA–CO–8 (IAEA-KST)	CaCO$_3$	−5.75 ± 0.06	Stichler, 1995
L–SVEC	Li$_2$CO$_3$	−46.48 ± 0.15	Stichler, 1995
IAEA–CO–9 (IAEA-NZCH)	BaCO$_3$	−47.12 ± 0.15	Stichler, 1995
USGS24	C (graphite)	−15.99 ± 0.11	Stichler, 1995
NBS 22	Oil	−29.74 ± 0.12	Gonfiantini et al., 1995; Coplen et al., 1983
RM 8452 (Sucrose ANU)	Sucrose	−10.43 ± 0.13	Gonfiantini et al., 1995
NGS1	CH$_4$ in natural gas	−29.0 ± 0.2	Hut, 1987
NGS1	C$_2$H$_6$ in natural gas	−26.0 ± 0.6	Hut, 1987
NGS1	C$_3$H$_8$ in natural gas	−20.8 ± 1	Hut, 1987
NGS2	CH$_4$ in natural gas	−44.7 ± 0.4	Hut, 1987
NGS2	C$_2$H$_6$ in natural gas	−31.7 ± 0.6	Hut, 1987
NGS2	C$_3$H$_8$ in natural gas	−25.5 ± 1	Hut, 1987
NGS2	CO$_2$ in natural gas	−8.2 ± 0.4	Hut, 1987
NGS3	CH$_4$ in natural gas	−72.7 ± 0.4	Hut, 1987
NGS3	C$_2$H$_6$ in natural gas	−55.6 ± 5	Hut, 1987
IAEA–CH–7 (PEF1)	Polyethylene	−31.83 ± 0.11	Gonfiantini et al., 1995

NOTE: Values for $\delta^{13}C$ given in per mille relative to VPDB, defined by assigning a $\delta^{13}C$ value of +1.95‰ to NBS 19 carbonate. (See Appendix E for information about obtaining these materials.)

sediment) that have small amounts of organic matter present in a complex mineral matrix.

Individual Organic Compounds

A wide range of high quality, non-certified carotenoid and chlorophyll chemical standards are commercially available (e.g., Sigma-Aldrich, DHI, and Roth). The availability of a mixed pigment reference standard and biological matrix reference materials would improve analytical performance in individual laboratories, facilitate method and laboratory in-

TABLE 4.4 Reference Materials Available to the Radiocarbon Community (Rozanski et al., 1992; Bryant et al., 2001)

Standard	Material	Consensus fm
IAEA C-1	Carbonate Carrara marble	0.0002
IAEA C-2	Carbonate freshwater travertine	0.4114
IAEA C-3	Cellulose 1989 growth of tree	1.2941
IAEA C-5	Subfossil wood E. WI forest	0.2305
IAEA C-6	ANU sucrose	1.5061
FIRI A, B	Wood	0.0033
FIRI C	Turbidite Carbonate	0.1041
FIRI D, F*	Dendro-dated wood (Belfast Scots Pine)	0.5705
FIRI E	St. Bees Head humic acid	0.2308
FIRI G, J	Barley mash	1.1069
FIRI H	Dendro-dated wood (Hohenheim Oak)	0.7574
FIRI I	Wood cellulose (Belfast Scots Pine)	0.5722
FIRI L*	Whalebone	0.2035

*No longer available

NOTE: The fraction modern (fm) listed is the consensus value reported by the directors of the studies. (See Appendix E for information about obtaining these materials.)

ter-comparisons, and help resolve inter-method variability during oceano-graphic expeditions. A variety of high quality, non-certified organic chemical standards are commercially available, e.g., from Alltech (fatty acids), Pierce Chemical Co. (amino acids), Sigma-Aldrich (phytol and a wide range of fatty acids and amino acids), Steraloids, Inc. (a wide range of marine and terrestrial sterols), and Chiron AS (diagenetic products of key lipid biomarkers, including deuterated internal standards).

Radionuclides

Several organizations (e.g., NIST, NRC-Canada, and IAEA) provide sediment reference materials containing radionuclides, many of which are only certified for artificial radionuclides (^{137}Cs, ^{90}Sr, ^{241}Am, and ^{239}Pu). Certain specific radionuclides have no certified natural matrix materials, including ocean, lake, and river sediments. Although these sediments are certified for a few naturally occurring and artificial radionuclides, the extent of radioactive equilibrium of the uranium and thorium decay series in these environmental materials is not provided. NIST currently offers an ocean sediment Standard Reference Material[3] (SRM 4357) in

---

[3]The term "Standard Reference Material" is a trademark of NIST.

which [230]Th, [226]Ra, [232]Th, [228]Th, and numerous artificial radionuclides are certified, and which give non-certified activities of uranium isotopes, ^{210}Pb, [228]Ra, and additional artificial radionuclides. This reference material is a blend of sediments collected from both the Chesapeake Bay and from the seafloor off of the British Nuclear Fuels Sellafield facility in the United Kingdom. The IAEA supplies reference materials in the form of a number of marine and terrestrial sediments, soils, and ores, some of which are certified for long-lived members of the uranium and thorium decay series, as well as trace elements. The IAEA reference materials are prepared in limited quantity and replaced frequently.

Radiochemists require reference materials with matrices that reflect sediments with widely different composition. For example recently deposited deep-sea sediments contain [230]Th and [231]Pa primarily as surface-adsorbed components, whereas sediments collected on (or close to) continental margins contain a higher proportion of structurally-bound components. No reference materials exist that can be used to test selective leaching procedures designed to identify how these radionuclides are incorporated into the matrix. Additionally, no reference materials exist with which to evaluate matrix-correction procedures in gamma-ray analyses or to compare analytical results of radionuclide measurements from different laboratories. To correct this problem, radiochemists need solid reference materials containing known amounts of the longer-lived radionuclides with an estimate of the degree of radioactive equilibrium through the series.

Reference materials that represent the primary deep-sea and coastal depositional environments and biological materials would solve many of the problems that radiochemists face in analysis of sediments from these settings. Radiochemists require reference materials comprising the primary end member sediment and biological types (calcium carbonate, opal, and red clay from the deep-sea and carbonate-rich, silicate-rich, and clay mineral-rich sediments from coastal environments and representative biological materials). Additional sediment reference material from a river delta would be valuable to test the release of radionuclides that occurs as riverine particles contact seawater.

RECOMMENDED REFERENCE MATERIALS

Given that analyses of particulate and sedimentary elements and compounds are highly matrix-dependent, the committee recommends that reference materials be made available for ten important matrix types, covering three biological and seven sedimentary forms described below, which also cover a large variety of diagenetic states. Rather than recommend that these reference materials be certified for a wide variety of constituents, the greatest value will be derived by producing stable, homogeneous materials that can be used to derive community consensus values for many specific analytes. Further certification could then be accomplished as needed.

Recommended Biological Matrices

The diversity of marine photosynthetic microalgae and bacteria is extensive, making it impractical to develop reference materials for all of the species currently in culture. For the three biological matrices, it is recommended that diatom (*Thalassiosira pseudonana*), dinoflagellate (*Scrippsiella trochoidea*), and haptophyte (*Emiliania huxleyi*) mass cultures be developed as reference materials in order to improve analytical performance and facilitate inter-laboratory comparison. These three organisms provide a wide range of oceanographically relevant mineral, trace metal, and organic analytes for the establishment of a comprehensive list of consensus values (Table 4.5). They represent three major matrices—opal, carbonate, and organic matter. They also include the two mineral phases thought to be important "ballast" materials for facilitating preservation and vertical export of POM to the sediments (Armstrong et al., 2002). Consensus analyte values (e.g., isotope ratios) can be obtained for both the inorganic (calcium carbonate, opal) and organic (POC, PN, and biomarkers) phases of these biological matrices. Biological reference materials can also be used for the preparation of both $C_{37:2}$ and $C_{37:3}$ alkenone standards (isolated from the *E. huxleyi*) and a mixed pigment standard containing a wide range of individual chlorophylls and carotenoids (prepared by mixing acetone extracts obtained from the three phytoplankton reference materials).

Recommended Sedimentary Matrices

Recommended sedimentary reference materials include three separate carbonate-, opal- and clay mineral- rich open-ocean sediments; three coastal sediments containing the same three mineral types; and a deltaic sediment that has not been in contact with seawater. To include relatively young material, each of these seven samples should be taken from within the top 10 cm of the sedimentary column. Table 4.6 suggests possible locations to sample the types of material matrices of interest. The combination of both coastal and open-ocean sediments would provide a wide range of diagenetic states, with the previously described phytoplankton reference materials providing fresh organic matrices and open-ocean red-clay material providing some of the most degraded sedimentary material that exists in the marine environment. The mineral types represented in

TABLE 4.5 Distribution of Mineral, Trace Metal, and Organic Analytes in the Target Biological Matrices Recommended as Reference Materials (+ = present and − = absent).

Analyte	T. pseudonana	S. trochoidea	E. huxleyi
Particulate Organic Carbon	+	+	+
Particulate Nitrogen	+	+	+
Minerals			
CaCO$_3$	−	−	+
Opal	+	−	−
Trace metals			
Fe	+	+	+
Zn	+	+	+
Tetrapyrroles			
Monovinyl chlorophyll a	+	+	+
Chlorophyll c_1	+	−	−
Chlorophyll c_2	+	+	+
Chlorophyll c_3	−	−	+
Carotenoids			
Dinoxanthin	−	+	−
Peridinin	−	+	−
Fucoxanthin	+	−	−
19′-hexanoyloxyfucoxanthin	−	−	+
Diadinoxanthin	+	+	+
Diatoxanthin	+	+	+
β,β-carotene	+	+	+
Phytol	+	+	+
Sterols	+	+	+
C$_{37-39}$ alkenones	−	−	+
Fatty acids	+	+	+
Amino acids	+	+	+
Nucleic acids (RNA and DNA)	+	+	+
Carbohydrates	+	+	+
Cellulose	−	+	−

TABLE 4.6. Examples of Locations Where the Recommended Sedimentary Reference Materials Could be Obtained

Matrix Types	Open-Ocean Locations	Coastal Locations
Carbonate	Mid-Atlantic Ridge	Florida Bay
Opal	Southern Ocean	Peru Upwelling
Clastic detrital	North Pacific Gyre	Gulf of Mexico
Deltaic clastic detrital		Atchafalaya River

Table 4.6 (opal, carbonate and aluminosilicate) were chosen to provide end members for matrix analysis. These samples also could be blended to produce any mixture desired. Sampling sediments from well-studied depositional regions such as MANOP Site R in the North Pacific (for open-ocean red clay) and Florida Bay (for carbonate sediments) would offer a variety of supplemental information that is already available. In addition, the (Atchafalaya) river sediment would provide a useful link between those studying terrestrial and marine processes.

As riverine particulate matter encounters seawater, a variety of changes may occur to the terrestrial particles. Some surface-bound ions are released to solution as they exchange with dissolved species. Such exchange reactions are especially important in understanding the geochemistry of radium. Evaluation of these exchange reactions is often approximated by collecting sediment from the freshwater region of an estuary and exposing it to seawater. However, such experiments may contain artifacts due to a lack of control of the simulated conditions. A freshwater deltaic reference sediment could be employed to help refine these experiments and eliminate artifacts.

POTENTIAL LONG-TERM NEEDS FOR ADDITIONAL REFERENCE MATERIALS

Some primary standards exist for the uranium to thorium series isotopes currently used extensively in the field. There is a need, however, to produce a certified reference material for uranium and thorium decay series isotopes with masses greater than ^{226}Ra, using a natural material such as uraninite. Sediment dating/mixing studies would benefit from a marine sediment reference material containing ^{210}Pb. The committee recognizes that producing this material in different matrices would be difficult, though not impossible. Few samples contain background levels for the cosmogenic and bomb-produced radionuclides ^{10}Be, ^{36}Cl, ^{26}Al,

^{241}Am, ^{90}Sr, ^{239}Pu, ^{240}Pu, ^{137}Cs, ^{129}I, or ^{14}C, which also lack solution standards. NIST has an active program to address the development of all of these radionuclide standards except ^{14}C.

Long-term needs for organic reference materials include ether-linked lipids, hopanoids, pheopigments, and mycosporine-like amino acids (MAA). Members of the *Archaea* possess diagnostic ether-linked isoprenoid lipids that are used in the construction of monolayer or bilayer cell membranes. These ether-linked lipids occur as glycerol diethers or diglycerol tetraethers. Hopanoids are chemical components found in bacterial cell membranes and are thought to play an important role in membrane stabilization. Ether-linked lipids and hopanoids are useful biomarkers for detecting the presence of *Archaea* and *Bacteria* in environmental samples. Pheopigments are magnesium-free chlorophyll degradation products that are useful as tracers of heterotrophic processes such as zooplankton grazing. MAAs are imino-carbonyl derivatives of mycosporine cyclohexenone. They are structurally diverse and widely distributed among marine cyanobacteria, microalgae, invertebrates and vertebrates. MAAs absorb strongly in the 310-360 nm waveband and are thought to protect marine organisms from the damaging effects of ultraviolet radiation. It is likely that the need for quantification of these organic compounds will grow in the future. Consequently, future matrix-based reference materials should be prepared from materials that contain these analytes.

5

Production and Distribution of Chemical Reference Materials

INTRODUCTION

The preceding chapters detail the extensive need for reference materials in the ocean sciences. This chapter focuses on what is required to produce new reference materials as well as how to encourage the use of those that already exist.

Many of the reference materials currently available do not represent the matrices in which oceanographers work, nor do they contain the desired analytes at the concentrations found in nature. Furthermore, the sources of reference materials and certified reference materials are diverse and information about them is not well-coordinated. Better cooperation and linkages amongst reference materials users and certified reference material producers should be encouraged. COMAR (http:// www.bam.de/cgi-bin/crm_office.cgi), an international database for sharing information between reference material suppliers is a good start but it is not "user-friendly," and its existence is not widely known within the oceanographic community. More must be done to share information and encourage production and use of reference materials and certified reference materials. Scientific societies and international organizations have a clear and present opportunity and responsibility to encourage training in analytical quality control, and the provision and use of reference materials in ocean science, thus enhancing the quality of measurements made.

REQUIREMENTS OF REFERENCE MATERIALS

Reference materials must fulfill certain rigorous criteria before they are accepted and found useful by the analytical community. The following conditions are prerequisites for preparing reference materials that are mutually acceptable to organizations around the world:

1. *Homogeneity.* Homogeneity assures that the analysis of all sub-samples of the reference material taken for measurement will produce the same analytical result within the stated measurement uncertainty. This is particularly important in the case of certified reference materials. Reference material producers therefore must specify the minimum amount of sample for which homogeneity has been measured and is valid. Finally, the ease of re-homogenizing the material after packaging must be taken into consideration.

2. *Stability.* Producers must state the length of the reference material's useable life, since they can be sensitive to light, humidity, microbial activity, temperature, time, etc. Long-term testing is required to validate the stability of a material under a variety of storage and transport conditions.

3. *Similarity to the real sample.* To produce meaningful analytical results, the reference material should mimic as closely as possible the matrix of the test sample.

4. *Accuracy, uncertainty, and traceability.* A certified value is the best approximation of the true concentration of the analyte. During the certification process, a variety of analytical methods may be used to determine this true value. Uncertainty estimates ultimately based on this process, together with information about the material's homogeneity can give a certified reference material traceability, needed for true international comparability.

REFERENCE MATERIAL AND CERTIFIED REFERENCE MATERIAL PRODUCTION

Recently, several books have been written that describe the requirements and protocols for the production of reference materials in general and for environmental science in particular. Due to space limitations, only a brief summary of these ideas is presented below. The reader is encouraged to consult these works directly if a more detailed discussion is required.

Zschunke (2000) has collected a series of general interest papers on the use of reference materials in analytical chemistry. While written for chemists, the book addresses both the certification process and the application of reference materials to a variety of applications from materials testing to environmental analysis. A brief summary of international col-

---

Box 5.1
General Steps in Preparation of a Reference Material
(Stoeppler et al., 2001)

1. Define the reference material, including the matrix, the properties to be certified, and their desired levels
2. Design sampling procedure
3. Design sample preparation procedure
4. Select a method appropriate for homogeneity and stability testing
5. Design the characterization of the reference material
6. Acquire samples
7. Prepare samples
8. Test for homogeneity
9. Test for stability
10. Characterize the reference material
11. Combine the results from homogeneity testing, stability testing, and characterization and assemble an uncertainty statement
12. Set up a certificate and, if appropriate, a certification report

---

laboration via the COMAR database is also included. For a discussion of the various steps of the certification process and the requirements for reference materials for a wide variety of specific substances see Stoeppler et al. (2001). Finally for a discussion of the preparation of matrix reference materials for various environmental applications, Parkany and Fajgelj (1999) provide a thorough discussion of example materials from several national metrology organizations along with chapters detailing their intended uses.

The preparation of a reference material requires substantial planning prior to undertaking a specific project (see Box 5.1). The process begins with the definition of the material to be produced, for example, "preparation of a seawater-based reference material containing the nutrient elements: NO_3, PO_4, and $Si(OH)_4$ at concentration levels appropriate to oceanic samples and certified for these constituents." Such definitions arise either from internal decisions by reference material producers (such as NIST or NRC-Canada) typically in response to perceived needs, or through external pressure on these producers from potential users. (This report, for example, explicitly identifies a number of pressing needs for reference materials for the ocean sciences.)

At the same time, decisions must be made regarding the producers of these materials. For new reference material needs, identified within agencies such as NIST or NRC-Canada, this entails a decision as to whether to produce the material in-house or to sub-contract the work elsewhere. For externally recognized needs, both a supplier (e.g., either one of the recognized national facilities or, possibly, an ad hoc group of investigators [see below]) and a funding source must be identified.

The next task is usually to obtain a sufficient amount of raw material with the desired properties. The amount of material needed depends on:

- the number of samples of reference material needed,
- the need for a feasibility study,
- the number of samples needed for the homogeneity study,
- the number of samples needed for the stability study, and
- the number of samples needed for characterization of the reference material (Stoeppler et al., 2001).

Once a sample is acquired, the supplier can begin to prepare the material in the state it will be used. It is particularly important at this stage to stabilize the material (if required) to prevent changes in composition of critical components and to homogenize the material so that all future sub-samples will be as identical as possible. If additional procedures (e.g., reduction of grain size) are needed to prepare the desired material, they should occur at this stage. At this point a preliminary assessment of homogenization should also be made.

Since one of the main goals of reference material production is to provide a *stable* reference material, tests for stability begin early in the production process. Ideally, these should be conducted over the expected lifetime of the reference material prior to its distribution; however, these tests can be conducted concurrently if required.

Once the stock material has been determined to be stable and homogeneous, it can be divided into portions appropriate for use as a laboratory reference material. The portion size depends on the analyst's expectations. In some cases, providing single-use samples is appropriate: the material may have so little stability after being opened that it is unlikely that it could be used again. In such cases, however, insufficient quantity can lead to low detection levels, thus increasing error. Alternatively, however, a single package can produce several replicates so that too much material leads to extensive re-use of the opened container. This can result in problems with reference material integrity and thus cause potentially inaccurate results.

Once the material has been packaged, it must then be checked for uniformity. Any additional characterization of the material, in the form in which it will be distributed (e.g., grain size, color) should be conducted at this stage. Subsequently, the material is analyzed for the analyte(s) of interest for certification purposes, after which the certificate of analysis can be prepared. It is also imperative that a certified reference material or reference material be continually monitored for stability throughout its useable lifetime. This process should also include periodic re-assessment of the certified concentration(s).

The useable lifetime of a reference material depends on both its inherent stability and also on its rate of use. It will thus be necessary to establish a timetable for the preparation of further batches of the material so as to ensure that supplies are always available. These subsequent batches (particularly if they are matrix-based) will need to be treated almost like a new material, i.e., they should be tested for stability, and a complete new certification will be needed for each replacement batch. This involves substantial work each time it is done, hence it is desirable to make as large a batch as practical, consistent with the probable shelf-life and expected usage.

METHODS EMPLOYED TO CHARACTERIZE REFERENCE MATERIALS AND CERTIFIED REFERENCE MATERIALS

There are several accepted methods for characterizing and producing reference values of reference materials and certified values of certified reference materials. The more widely accepted methods include:

1. *Certification using one definitive method.* This option is employed when a highly established, internationally accepted scientific primary method is available. The method must be shown to have negligible systematic errors and to provide sufficient measurement accuracy. An example of such a method is isotope dilution mass spectrometry.

2. *Certification through interlaboratory testing.* In this case, the reference/certified value is obtained by pooling results from several laboratories that have demonstrated capability in analyzing the analyte(s) of interest. The various laboratory means are manipulated statistically to determine the best or truest estimate of the value of interest.

3. *Certification using at least two independent methods.* At least two validated, robust, and independent methods are employed to produce the true value of the analyte.

Often, combinations of the above methods are used to certify reference materials. For example, the two-independent method approach is often coupled with the inter-laboratory testing procedure, after which the data are combined to give the proposed best value.

PREPARATION OF RECOMMENDED NEW REFERENCE MATERIALS AND CERTIFIED REFERENCE MATERIALS FOR OCEAN SCIENCE

The rationale for choosing particular reference materials (and, in some cases, for recommending certification of reference materials) was discussed in Chapters 3 and 4. In this section, the committee proposes specifications for such reference materials, as well as some further suggestions specific to the preparation of the recommended reference materials.

Most of the reference materials discussed are based on natural matrices (seawater, algal cells, sediment) and would initially only be certified for a limited number of constituents. Nevertheless, it is apparent that such materials provide a resource for the investigation of a much wider variety of constituents, and it is important that the ocean science community be encouraged to investigate these materials further. In particular, the existence of these materials would facilitate a wide variety of necessary interlaboratory method comparisons that have been neglected to date. Eventually these intercomparisons will result in consensus values for other constituents, which can then be assigned to the reference materials.

Approaches to the Preparation of Seawater Nutrient Reference Material(s)

The preservation of nutrient solutions at the concentrations occurring in natural seawater is a major challenge to the routine production of a nutrient reference material. Preservation techniques must be developed that maintain concentrations stable for periods of at least one to two years. Gamma radiation will produce nitrite that is unstable. Therefore this method appears to be problematic. The feasibility of other techniques, such as autoclaving, ultra-violet or microwave radiation, freezing, and acidification, should be evaluated.

A single technique that preserves all three nutrients (NO_3, PO_4, and $Si(OH)_4$) in a single matrix would be most desirable. If this proves impossible, it will be necessary to use different approaches to preserve and store different nutrients. For example, glass containers cannot be used to store a reference material for $Si(OH)_4$ due to the slow dissolution of silica (Zhang et al., 1999), while the polymerization of silicic acid upon freezing eliminates that option for preserving stable $Si(OH)_4$ levels (Zhang and Ortner, 1998).

Nutrient calibration solutions in seawater are commonly prepared by dissolving known amounts of pre-dried, solid, primary standard salts in low-nutrient seawater. Low-nutrient seawater must be collected from oligotrophic open-ocean surface water to minimize background nutrient contents. Such samples could be obtained from the Sargasso Sea during the summer after spring phytoplankton blooms have depleted most of the surface nutrients. Obtaining sufficiently low-nutrient seawater might require further biological depletion or chemical removal following collection. Quantification of trace nutrients in collected seawater (or at least demonstration that they are below the detection limits of most methods) will be necessary to obtain the high accuracy required for a certified reference material.

Approaches to the Preparation of Seawater Trace Metal Reference Materials

The only element whose concentration is recommended to be certified is iron. An iron reference material will also clearly be useful for studying other important metals such as zinc, manganese, copper, molybdenum, cobalt, vanadium, lead, aluminum, cadmium, and the rare earth elements. It is thus desirable to assure stability for some of these elements in addition to iron.

The optimal size for a portion of a trace metal reference material is 1L, so large volumes (greater than 1000 L) must be collected and prepared, which is feasible given currently available Teflon® storage containers, sampling systems, and pumps. More experience is required, however, in collecting large volumes of uncontaminated water.

Preservation is a critical issue that has been carefully studied in the past. Acidifying materials with HCl to a pH of 2.3 (by adding 7.5 mmol acid to each liter of seawater) provides adequate preservation against precipitation or wall loss for most metals. Low density polyethylene bottles are currently considered to be acceptable for storage of acidified oceanic seawater. However, they require careful cleaning using hot surfactants to remove organic softeners and contaminants, as well as acids. They also require copious rinsing to remove contaminating metals. A pH of 2.3 is recommended for acidified seawater as a trade-off between stabilizing the sample and minimizing the aggressiveness of the stored seawater and the need for neutralizing the sample pH prior to analysis.

At present, isotope-dilution mass spectrometry provides the best method to certify iron concentration in the recommended deep-water reference material (an expected iron concentration of approximately 0.7 nM) and to obtain an information value for iron in the recommended surface water reference material (an expected iron concentration of approximately 50 pM or less).

Approaches to the Preparation of
Radionuclide Reference Material Solutions

Radionuclide reference materials can be prepared from natural ores of uranium and thorium. The uranium ore selected should have concordant ages, indicating that the system has been closed long enough to establish radioactive equilibrium through the long-lived nuclides. Specific measurements of uranium, thorium, protactinium, actinium, and radium isotopes are required to ensure that radioactive equilibrium is still present. The establishment of radioactive equilibrium for the thorium series takes only about 35 years, so thorium salts purified 35 years ago could be used for this solution. It is necessary to measure contamination from uranium series daughters in the thorium solution. For the ^{210}Pb / ^{210}Po reference material, the solution could be prepared from ^{210}Pb that is at least two years old. It is important that there be no ^{226}Ra in this solution, otherwise the ^{210}Pb activity would not change in a predictable manner due to variable escape of ^{222}Rn.

Approaches to the Preparation of Algal-Based Reference Materials

Each of the algal based reference materials recommended in Chapter 4 should be certified for both inorganic and organic carbon concentrations, total nitrogen concentration, δ^{13}C of both the inorganic and the organic carbon components, and δ^{15}N for the total nitrogen component.

These three biological matrices can be prepared by individually growing the *T. pseudonana*, *S. trochoidea* and *E. huxleyi* clones in large-volume photobioreactors, such as those available at NRC-Canada's Institute for Marine Biosciences. Because growth conditions (light, nutrients, and temperature) can greatly affect the composition of phytoplankton, the photobioreactors must be tightly regulated to ensure that reproducible harvests of phytoplankton can be produced over time. Once harvested, the algal material can be freeze-dried and packaged for subsequent testing and distribution. Specific clones of cryogenically preserved microalgae should be obtained from a suitable source (e.g., Center for Culture of Marine Phytoplankton, Bigelow Laboratory for Ocean Sciences, W. Boothbay Harbor, ME) and used to inoculate the photobioreactors in order to minimize batch-to-batch variability caused by genetic drift in stock cultures.

Approaches to Preparation of Sediment-Based Reference Materials

Each of the sediment based reference materials recommended in Chapter 4 should be certified for both inorganic and organic carbon concentration, total nitrogen concentration, δ^{13}C of both the inorganic and the organic carbon components, and $\delta^{15}N$ for the total nitrogen component. These matrix types (opal, carbonate, and aluminosilicate) provide end members for analysis and could be blended to produce any mixture desired. Sediment materials collected at the locations in Table 4.6 should be taken from within the top 10 cm of the sediment. Large box cores are the most appropriate for collecting a sufficiently large sample. The sedimentary reference materials could best be made available in a freeze-dried form, with known salt content, homogeneous at the mg level. To achieve homogeneity, sediments could first be ground and sieved through a plastic sieve to 42 μm, then mixed and either bottled or canned. Some unprocessed frozen sample could also be archived.

COSTS OF PRODUCING AND DISTRIBUTING REFERENCE MATERIALS AND CERTIFIED REFERENCE MATERIALS

Although a detailed discussion of the economics of producing reference materials for the ocean sciences is beyond the purview of this committee, it was felt appropriate to indicate here the principal factors that influence the overall cost of a particular material.

As has been discussed, several steps are required to prepare reference materials. Each step in the production process carries associated costs that will vary according to the complexity required to produce the final material. Given the additional requirements of traceability and the rigorous certification criteria, a certified reference material can be much more costly to produce than simple reference and/or consensus materials. Although the exact cost of certified reference materials and reference materials are specific to each individual material, particular issues common to all materials should be considered.

The cost of developing and producing certified reference materials, especially matrix materials, is high. Researchers should realize, however, that in most cases the true cost of a certified reference material is never fully charged to the customer. For example, the costs of the research required to demonstrate the feasibility of producing a certified reference material are usually not incorporated into the final cost. As a result, a commercial organization may never want to undertake this initial research, since profits would often not be realized.

The cost of a certified reference material is influenced not only by the cost of its intensive certification procedure but also by collection, preparation, distribution, storage, stability assessment, the required annual or semi-annual reassessment, and general production costs. Organizations offering after-sales technical support bear additional personnel costs. To keep costs down, organizations producing certified reference materials have been trying to produce larger batches so that the various costs are amortized over a larger number of samples. However, stability concerns may not always make larger batches a feasible option, particularly if the number of portions produced would not be distributed in an appropriate time frame.

A STRATEGY FOR THE PRODUCTION OF NEW REFERENCE MATERIALS FOR THE OCEAN SCIENCES

Although the scientific returns made possible by the existence of well-characterized reference materials are clear, it is also apparent that the traditional approach in which reference materials are developed, prepared, and distributed by national standards organizations (such as NIST) provides limited value to the ocean sciences. Such organizations have many demands on their resources, and typically will not be able to place a high priority on the development, production, and distribution of reference materials for the ocean science research community in particular.

An alternate approach, discussed at length by the committee, directly involves the oceanographic research community in the process of development, production, and distribution of reference materials—ideally in partnerships with experienced reference material producers such as NIST or NRC-Canada. Appropriate science funding agencies could fund initial development costs, in partnership with various government and private "standards" producers. For example, in response to the needs of the National Status and Trends Program, NOAA contributed funds to the production of eight NIST Standard Reference Materials and seven calibration solutions. The SRMs were based on natural matrices and are prepared at two concentration levels. The calibration solutions are for each of the three chemical classes of analytes quantified by the National Status and Trends Program. These solutions were prepared at the request of the National Status and Trends contract laboratories and are currently available for purchase from NIST. In addition, the National Status and Trends Program also contracts with NRC-Canada to prepare unknown samples suitable for use in laboratory intercomparison exercises.

Direct participation of the oceanographic research community will enable collection of matrix-based materials such as natural seawater or sediments in an efficient manner (for example, material collection could be combined with other scheduled research activities). Furthermore, the extensive analytical skills of the oceanographic research community can be harnessed to conduct the necessary testing for homogeneity, stability, and ultimately characterization of the proposed reference materials. At the same time, complete success requires the participation of reference material producers, who bring to the table extensive knowledge and experience in the preparation and stabilization of reference materials, and on occasion access to necessary facilities.

This collaborative approach to the development of consensus materials could thus result in the eventual certification of those materials, if required. Arrangements could possibly be developed with existing certified reference material or reference material producers to disseminate information and to distribute materials to the oceanographic research community.

The development and initial production of any new reference material would require initial funding. Proceeds from the sales of the material could then be used towards the production of replacement materials. The full cost of the production and distribution of the material might be recovered in time, particularly for widely-used materials for which no further development is needed; the research activities using these reference materials would then bear future costs directly.

EDUCATION

Funding agencies must commit to encouraging and funding both the development and distribution of reference materials and certified reference materials. However, production and distribution of reference materials will not become self-supporting unless the ocean research community uses them and widely appreciates their value. At present, this is not the case; therefore, any plan to produce reference materials must also be designed to encourage their use.

The first step in such a plan is education. Training in the field of oceanography needs to place greater emphasis on analytical quality control, particularly with respect to accuracy to complement the current focus on precision. For those already established in the field (and perhaps entrenched in their ways), national and international meetings should incorporate training courses and workshops.

Participation in round-robin exercises offers a substantial impetus for improvements in analytical quality control within individual laboratories. During these events, participating laboratories individually analyze samples of a particular test material. Such exercises must be organized using materials and analytes relevant to the ocean sciences. Laboratories must be encouraged to participate, even if they are at an early stage in their experience with the relevant analytical techniques.

Those scientists using available reference materials should be encouraged to report such uses explicitly in the scientific literature. A recent article by Jenks and Stoeppler (2001) goes so far as to suggest that scientific publishers should provide explicit recommendations as to how and where in a paper the use of certified reference materials should be described. Proposal and journal article reviewers also need to be encouraged to question the analytical quality control (and ultimate value) of measurements made without the benefit of reference materials.

6

Conclusions and Recommendations

Substantially increased awareness and availability of reference materials offers many benefits to the ocean sciences. The regular use of such materials can provide a much-needed basis for interlaboratory and international comparison of results, making it possible to acquire accurate, meaningful global data sets that can be used to study problems requiring observations on large space and time scales. However, reference materials are costly to produce—particularly if they are certified for a number of constituents—and it has not always been clear to the ocean science community that this cost will be repaid with significant added value.

In the case of the acquisition of large-scale data sets, the benefits of using reference materials are self-evident. In the past, whenever such data sets have been acquired without using suitable reference materials, a great deal of effort has subsequently been needed to adjust the data to a common scale. But the benefit of comparability is not restricted to large programs. Matrix-based reference materials that can be exchanged between different laboratories will enable researchers to better understand their own techniques and the information they provide.

Seawater studies require certified reference materials for biologically important dissolved components such as carbon (both inorganic and organic), nutrients, and trace metals, as well as for salinity, which is hydrographically important. A number of the committee's key recommendations therefore explicitly address these parameters. There is also a striking need for reference materials based on particulate matrices, where many of the analytical techniques used are matrix dependent and differ markedly in their implementation from laboratory to laboratory. In response, the committee proposes both the preparation of such materials as well as the development of a vision for community involvement in reference material production, characterization, and use (see Chapter 5). This involvement will ultimately create a better understanding of the future needs for certified reference materials.

Any new effort to provide reference materials to the ocean sciences community must also put in place a number of educational and advertising efforts to make researchers aware of the existence of appropriate reference materials and to provide instruction on how to make best use of what is bound to be, at best, a finite resource. The committee, therefore, also identifies education as a critical need.

RECOMMENDATIONS FOR REFERENCE MATERIALS FOR OCEAN SCIENCE

The development of new reference materials should not be undertaken lightly. Producing the prototype of each new material—including testing of the material for homogeneity and stability—requires substantial investment, and still further investment is required to provide accurate analyses of certain properties (as discussed in Chapter 5). Reference materials recommended in this report have been selected from a wide range of possible candidates. Their availability to the ocean science community will enable meaningful progress across a range of scientific questions.

Presently Available Reference Materials

In the past, a limited number of reference materials have been explicitly developed for ocean science (as discussed in Chapter 2): salinity, ocean CO_2, and DOC. Although salinity reference materials are available on a commercial basis from Ocean Scientific International Ltd. in the United Kingdom, the others are presently supported through grants from the U.S. National Science Foundation. The widespread use of such materials and their success in enhancing the scientific return on ocean studies is clear, and it is essential that such materials remain available.

In addition, NIST presently prepares a number of standard reference materials that are of immediate use to the ocean science community. These include materials for ^{14}C (SRM 4990C) and ^3H (SRM 4361C) as well as ^{238}U, ^{234}U, ^{235}U (SRM 4321C), ^{230}Th (SRM 4342—presently out of stock), ^{226}Ra (SRM 4969), ^{228}Ra (SRM 4339B), ^{10}Be (SRM 4325), ocean sediment (SRM 4357), and river sediment (SRM 4350B). It is important to assure the continued availability of these materials.

Recommended New Reference Materials for Seawater Studies

Recommended Nutrient Reference Materials

The committee recommends the development of a seawater-based reference material containing the nutrient elements: nitrogen (as NO_3), phosphorus (as PO_4), and silicon (as $Si(OH)_4$) at concentrations similar to those in oceanic deep waters (40 μM for NO_3, 3 μM for PO_4, and 150 μM for $Si(OH)_4$) and certified for these constituents.

There is an urgent need for a certified reference material for these nutrients. Completed global surveys already suffer from the lack of previously available standards, and the success of future surveys as well as the development of instruments capable of remote time-series measurements will rest on the availability and use of good nutrient reference materials.

Recommended Trace Metal Reference Materials

The committee recommends the development of two reference materials for seawater trace metal analysis:

- A seawater-based reference material with concentrations of metals corresponding to oceanic deep water, certified for total iron concentration.
- A seawater-based reference material with concentration of metals corresponding to open ocean surface water, with an information value for total iron concentration.

A significant proportion of the needs for reference materials for seawater trace metal studies would be addressed by the preparation of these materials. Although the total iron concentration of these reference materials should be provided, these materials clearly will be useful for studies of other important metals such as: zinc, manganese, copper, molybdenum, cobalt, vanadium, lead, aluminum, cadmium, and the rare earth elements. With careful planning, such water samples should be useful for analysis of dissolved organic substances as well. The collection sites should be chosen carefully to provide both a high and a low concentration reference material for as many metals as possible.

The committee recommends, but assigns a lower priority to, the preparation of reference materials from other locations. For example a standard for dissolved iron in a coastal seawater matrix containing high concentrations of dissolved organic material would be particularly useful in addressing matrix effects associated with such materials.

Recommended Radionuclide Reference Materials

The committee recommends the development of three radionuclide reference solutions:

- A synthetic certified reference material based on an acidic solution containing ^{238}U and ^{235}U (ca. 20 Bq/g and ca. 1 Bq/g respectively) with daughters in secular equilibrium through ^{226}Ra and ^{223}Ra.
- A synthetic certified reference material based on an acidic solution containing ^{232}Th (ca. 20 Bq/g) with daughters in secular equilibrium through ^{224}Ra.
- A synthetic certified reference material based on an acidic solution containing ^{210}Pb (ca 20 Bq/g) with daughters in secular equilibrium through ^{210}Po.

The presence of ^{222}Rn in the ^{238}U series makes the extension of the series through ^{210}Pb very difficult due to the escape of radon gas (see Chapter 5 for further preparation details).

Recommended Solid Reference Materials

Solid matrix-based reference materials are recommended for ten important matrix types: three based on algal materials and seven on sediments (details of possible preparation techniques are given in Chapters 4 and 5). Furthermore, the committee recommends that each of these materials be certified for both inorganic and organic carbon concentrations, total nitrogen concentration, $\delta^{13}C$ of both the inorganic and the organic carbon components, and $\delta^{15}N$ for the total nitrogen component. These are fundamental parameters whose measurements are not yet fully agreed upon.

All of these sediment reference materials would provide stable homogeneous materials containing a wide variety of chemical constituents that could be studied at the discretion of the ocean sciences community, and may ultimately be assigned consensus values for a number of important additional organic and inorganic analytes.

Recommended Algal-based Reference Materials

The committee recommends the development of three types of algal-based reference materials:

- A reference material based on a freeze-dried culture of the diatom *Thalassiosira pseudonana*.

- A reference material based on a freeze-dried culture of the dinoflagellate *Scrippsiella trochoidea.*
- A reference material based on a freeze-dried culture of the haptophyte *Emiliania huxleyi.*

These three cultures represent three major biogenic matrices: opal, carbonate, and organic matter, thus providing materials that can be used to investigate a variety of matrix-dependent effects (Table 4.5).

Recommended Sediment-based Reference Materials

The committee recommends seven types of sediment-based reference materials:

- Open-ocean, carbonate-rich, sediment-based reference material.
- Open-ocean, silicate-rich, sediment-based reference material.
- Open-ocean, clay mineral-rich, sediment-based reference material.
- Coastal, carbonate-rich, sediment-based reference material.
- Coastal, silicate-rich, sediment-based reference material.
- Coastal, clay mineral-rich, sediment-based reference material.
- Deltaic sediment-based reference material (that has not contacted seawater).

Table 4.6 lists locations where such sediment types can be found. Taken together with the algal-based materials, the coastal and open-ocean sediments will provide material encompassing a wide range of early diagenetic states. The algal materials provide a fresh organic matrix, while the open-ocean red-clay material represents some of the most degraded sedimentary material that exists in the marine environment. The other proposed sediments lie between these two extremes. These matrix types (opal, carbonate, and aluminosilicate) provide end members for analysis and could be blended to produce any mixture desired.

Summary

The various reference materials described above are listed together in Box 6.1. Each of these is considered by the committee to have the potential for significant impact on an important area of ocean science and, as such, is assigned a high priority (however, see Statement of Top Priorities below).

Box 6.1
Recommended Reference Materials for Ocean Science

Materials Recommended for Continued Availability:

Currently available materials:
1. Standard Seawater (*Ocean Scientific International Ltd.*)
2. Reference Materials for Ocean CO_2 (*NSF via Dr. A. Dickson*)
3. Reference Materials for Dissolved Organic Carbon (*NSF via Dr. D. Hansell*)
4. Various Standard Reference Materials from NIST (see text)

Materials Recommended for Development:

Seawater-based reference materials:
5. One certified for the nutrient elements: nitrogen (as NO_3), phosphorus (as PO_4), and silicon (as $Si(OH)_4$).
6. One with concentrations of metals corresponding to oceanic deep water, certified for total iron concentration.
7. One with concentrations of metals corresponding to open-ocean surface water with an information value for total iron concentration.

Certified reference materials for radionuclides:
8. An acidic solution containing ^{238}U and ^{235}U with daughters in secular equilibrium through ^{226}Ra and ^{223}Ra.
9. An acidic solution containing ^{232}Th with daughters in secular equilibrium through ^{224}Ra.
10. An acidic solution containing ^{210}Pb with daughters in secular equilibrium through ^{210}Po.

*Solid matrix-based reference materials:**
11. Freeze-dried culture of the diatom *Thalassiosira pseudonana*
12. Freeze-dried culture of the dinoflagellate *Scrippsiella trochoidea*
13. Freeze-dried culture of the haptophyte *Emiliania huxleyi*
14. Open-ocean, carbonate-rich sediment
15. Open-ocean, silicate-rich sediment
16. Open-ocean, clay mineral-rich sediment
17. Coastal, carbonate-rich sediment
18. Coastal, silicate-rich sediment
19. Coastal, clay mineral-rich sediment
20. Deltaic sediment (that has not contacted seawater)

---

*Each of these solid reference materials should be certified for both inorganic and organic carbon concentrations, total nitrogen concentration, $\delta^{13}C$ of both the inorganic and the organic carbon components, and $\delta^{15}N$ for the total nitrogen component.

RECOMMENDATIONS FOR COMMUNITY PARTICIPATION

In addition to making new reference materials available to the ocean sciences community, it is essential that a number of other strategies be put in place. These strategies will ensure optimal use of such materials, thus increasing their value to the ocean sciences and justifying this investment.

Quality Assurance and Quality Control of Large Projects

The use of appropriate reference materials should be a key feature of the quality assurance/quality control structure in any future ocean science project involving chemical measurement. Reference materials use should be explicitly addressed in the project planning stages, proposals, and publications. In the event that appropriate reference materials are not already available, the committee recommends that the proposed project develop a strategy for preparation to assure the ultimate value of measurements made as part of the research.

Database of Reference Material Availability

It is essential to develop and maintain a searchable, user-friendly database that ocean scientists can access to learn about those reference materials that are of particular interest to their research. A number of materials are currently available, but are rarely employed, in part because individual scientists may not be aware of their existence.

Courses on the Use of Reference Materials in Quality Control

Many ocean scientists may be unaware of the clear gains to be garnered by the appropriate use of reference materials, and of the best way to use such materials to ensure quality control of their analytical measurements. The best way to address these deficiencies is to encourage the presentation of short courses on this topic in conjunction with national meetings for ocean scientists. Ideally, such courses would be co-taught by individuals from both the ocean science and the reference material provider communities (such as NIST or NRC-Canada).

Encouragement of Participation in Round-Robin Exercises

A substantial impetus for improvements in analytical quality control within individual laboratories comes from participation in round-robin exercises in which participating laboratories independently analyze samples of a particular test material for specified analytes. It is important that round-robin exercises be organized using materials and analytes relevant to the ocean sciences and that laboratories be encouraged to participate, even at an early stage in their experience with the relevant analytical techniques.

Encourage Use of New Ocean Science Reference Materials

Most of the reference materials proposed in this report are based on natural matrices (seawater, algal cells, sediment) and will initially only be certified for a limited number of constituents. Nevertheless, it is apparent that such materials provide a resource for the investigation of a much wider variety of constituents, and it is important that the ocean science community be encouraged to investigate these materials further. In particular, these materials would enable a wide variety of necessary interlaboratory method comparisons that have been neglected to date, and ultimately additional consensus values for further constituents will be assigned to various research materials.

Enhanced Community Awareness

If the ocean sciences are to move forward and adopt the regular use of reference materials, the advantages of using such materials (and the pitfalls of not using them) need to be more broadly disseminated. Proposal and journal article reviewers should question the analytical quality control of measurements made without the benefit of reference materials. Outreach focused on how to use reference materials in the ocean sciences will further increase the awareness of the individual investigators and provide a focus for proper handling and application of a laboratory's internal reference materials.

Summary

The recommendations made in this section are listed together in Box 6.2.

STATEMENT OF TOP PRIORITIES

The committee recognizes that, by current standards, these recommendations will require a significant investment in reference materials for the ocean sciences. While such an investment is warranted, and the consequent gains to the ocean sciences will fully justify such expenditures, there is a need to assign additional priority to a subset of these materials to guide future planning.

Box 6.2
Additional Recommendations for Community
Participation

1. The use of appropriate reference materials should be a key feature of the quality assurance/quality control structure in any future ocean science project involving chemical measurement. Reference materials use should be explicitly addressed in the project planning stages, proposals, and publications.

2. It is essential to develop and maintain a searchable, user-friendly database that ocean scientists can access to learn about those reference materials that are of particular interest to their research.

3. It is essential to encourage the presentation of short courses on the best way to use reference materials to ensure quality control of analytical measurements in conjunction with national meetings for ocean scientists.

4. It is important that round-robin exercises be organized using materials and analytes relevant to the ocean sciences and that laboratories be encouraged to participate, even at an early stage in their experience with the relevant analytical techniques.

5. It is important that the ocean science community be encouraged to investigate the various proposed matrix-based reference materials so as to establish their properties with consensus-based values for the concentrations of a variety of constituents.

6. Proposal and journal article reviewers need to be encouraged to question the analytical quality control of measurements made without the benefit of reference materials.

The committee agreed that it was essential to ensure that reference materials for salinity, ocean CO_2, and DOC be kept available. These materials are now used regularly and are contributing to improvements in the science that can be achieved. The next highest priority is the development of a seawater-based nutrient reference material. Work on this material is presently in progress at NRC-Canada, and should be encouraged.

Two trace metal reference materials (one based on surface ocean water and one based on deep ocean water) are urgently needed to further the research community's ability to investigate the role of trace metals in ocean biogeochemistry. Thus these two materials are also assigned a high priority.

Although each of the ten materials proposed as solid reference materials provides clear benefits (each represents a unique matrix), the following subset was considered the highest priority: reference materials based on *Thalassiosira pseudonana* cells (providing an opal matrix), on *Emiliania huxleyi* (providing a carbonate matrix), and on the three open-ocean sediments (providing opal, carbonate, and aluminosilicate matrices). These priority materials should be initially certified for organic carbon and nitrogen.

To take advantage of these solid reference materials (and to a lesser extent the trace metal seawater-based materials), it will be essential to establish an infrastructure for the collation and redistribution of the information that is accumulated about these materials through community-wide use.

The final, and yet perhaps the highest priority of all, is to improve the awareness and knowledge base within the ocean science community as to the availability and optimal use of reference materials. To ensure quality control of measurements in ocean sciences, it is paramount to establish and maintain a database of reference materials, and to make available short courses in reference material use.

References Cited

Altabet, M.A., and R. Francois. 1994. Sedimentary nitrogen isotopic ratio as a recorder for surface ocean nitrate utilization. Global Biogeochemical Cycles 8(1): 103-116.

Aminot, A., and R. Keroul. 1991. Autoclaved seawater as a reference material for the determination of nitrate and phosphate in seawater. Analytical Chimica Acta 248:277-283.

Aminot, A., and R. Keroul. 1995. Reference material for nutrients in seawater: Stability of nitrate, nitrite, ammonia and phosphate in autoclaved samples. Marine Chemistry 49:221-232.

Anderson, L.A. 1995. On the hydrogen and oxygen content of marine phytoplankton. Deep-Sea Research I 42(9):1675-1680.

Armstrong, R.A., C. Lee, J.I. Hedges, S. Honjo, and S.G. Wakeham. 2002. A new, mechanistic model for organic carbon fluxes in the ocean based on the quantitative association of POC with ballast minerals. Deep-Sea Research II 49:219-236.

Bard, E., M. Arnold, B. Hamelin, N. Tisnerat-Laborde, and G. Cabioch. 1999. Radiocarbon calibration by means of mass spectrometric ^{230}Th/^{234}U and ^{14}C ages of corals: An updated database including samples from Barbados, Mururoa, and Tahiti. Radiocarbon 40(3):1085-1092.

Behrenfeld, M.J., J.T. Randerson, C.R. McClain, G.C. Feldman, S.O. Los, C.J. Tucker, P.G. Falkowski, C.B. Field, R. Frouin, W.E. Esaias, D.D. Kolber, and N.H. Pollack. 2001. Biospheric primary production during an ENSO transition. Science 291:2594-2597.

Béjà, O., L. Aravind, E.V Koonin, M.T. Suzuki, A. Hadd, L.P. Nguyen, S.B. Jovanovich, C.M. Gates, R.A. Feldman, J.L. Spudich, E.N. Spudich, and E.F. DeLong. 2000. Bacterial rhodopsin: Evidence for a new type of phototrophy in the sea. Science 289:1902-1906.

Benner, R. 2002. Chemical composition and reactivity. In Biogeochemistry of Marine Dissolved Organic Matter, D.A. Hansell and C.A. Carlson, eds., San Diego, California: Academic Press

Benner, R., J.D. Pakulski, M. McCarthy, J.I. Hedges, and P.G. Hatcher. 1992. Bulk chemical characteristics of dissolved organic matter in the ocean. Science 255:1561-1564.

Bergamaschi, B.A., J.S. Walters, and J.I. Hedges. 1999. Distribution of uronic acids and O-methyl sugars in sedimentary particles in two coastal marine environments. Geochimica et Cosmochimica Acta 63:413-425.

Bidigare, R.R., and C.C. Trees. 2000. HPLC phytoplankton pigments: Sampling, laboratory methods, and quality assurance procedures. Pp. 154-161 in Ocean Optics Protocols for Satellite Ocean Color Sensor Validation, Revision 2, J. Mueller and G. Fargion, eds., NASA Technical Memorandum 2000-209966.

Bidigare, R.R., A. Fluegge, K.H. Freeman, K.L. Hanson, J.M. Hayes, D. Hollander, J.P. Jasper, L. King, E.A. Laws, J. Milder, F.J. Millero, R.D. Pancost, B.N. Popp, P.A. Steinberg, and S.G. Wakeham. 1997. Consistent fractionation of ^{13}C in nature and in the laboratory: Growth rate effects in some haptophyte algae. Global Biogeochemical Cycles 11:279-292.

Bidigare, R.R., K.L. Hanson, K. Buesseler, S.G. Wakeham, K.H. Freeman, R.D. Pancost, F.J. Millero, P. Steinberg, B.N. Popp, M. Latasa, M.R. Landry, and E.A. Laws. 1999. Iron-stimulated changes in ^{13}C fractionation and export by equatorial Pacific phytoplankton: Toward a paleo-growth rate proxy. Paleoceanography 14: 589-595

Boaretto E, C. Bryant, I. Carmi, G. Cook, S. Gulliksen, D. Harkness, J. Heinemeier, J. McClure, E. McGee, P. Naysmith, G. Possnert, M. Scott, H. van der Plicht, M. van Strydonck. 2002. Summary findings of the Fourth International Radiocarbon Intercomparison (FIRI), 1998-2001. Journal of Quaternary Science 17(7).

Brassell, S.C. 1993. Applications of biomarkers for delineating marine paleoclimatic fluctuations during the Pleistocene. Pp. 699-738 in Organic Geochemistry, M.H. Engel and S.A. Macko, eds., New York: Plenum.

Brassell S.C., G. Eglinton, I.T. Marlowe, U. Pflaumann. and M. Sarntheim. 1986. Molecular stratigraphy: A new tool for climatic assessment. Nature 320:129-133.

Broecker, W. 1974. "NO" a conservative water-mass tracer. Earth and Planetary Science Letters 23:100-107.

Broecker, W.S., and T.-H. Peng. 1982. Tracers in the Sea. New York: Eldigo Press.

Broecker, W.S., M. Klas, E. Clark, G. Bonani, S. Ivey, and W. Wolfli. 1991. The influence of $CaCO_3$ dissolution on core top radiocarbon ages for deep-sea sediments. Paleoceanography 6(5):593-608.

Broecker, W.S., T.H. Peng, and T. Takahasi. 1980. A strategy for the use of bomb-produced radiocarbon as a tracer for the transport of fossil fuel CO_2 into the deep-sea source regions. Earth and Planetary Science Letters 49:463-468.

Browne, E. and R.B. Firestone. 1986. Table of Radioactive Isotopes. V.S. Shirley, ed., John Wiley and Sons, NY.

Bruland, K.W. 1980. Oceanographic distributions of cadmium, zinc, nickel, and copper in the North Pacific. Earth and Planetary Science Letters 47:176-198.

Bryant, C., I. Carmi, G.T. Cook, S. Gulliksen, D.D. Harkness, J. Heinemeier, E. McGee, P. Naysmith, G. Possnert, E.M. Scott, J. van der Plicht, M. van Strydonck. 2001. Is comparability of ^{14}C dates an issue? A status report on the Fourth International Radiocarbon Intercomparison. Radiocarbon 43:321-324.

Buesseler, K.O. 1991. Do upper-ocean sediment traps provide an accurate record of particle flux? Nature 353:420-423.

Carter, P.W., and R.M. Mitterer. 1978. Amino acid composition of organic matter associated with carbonate and non-carbonate sediments. Geochimica et Cosmochima Acta 58:1231-1238.

Charette, M.A., K.O. Buesseler, and J.E. Andrews. 2001. Utility of radium isotopes for evaluating the input and transport of groundwater-derived nitrogen to a Cape Cod estuary. Limnology and Oceanography 46:465-470.

Chavez, F.P., P.G. Strutton, G.E. Friederich, R.A. Feely, G.A. Feldman, D. Foley, and M.J. McPhaden. 1999. Biological and chemical response of the equatorial Pacific Ocean to the 1997 and 1998 El Niño. Science 286:2126-2131.

Cheng-Tung, A.C., L. Chi-Ming, H. Being-Ta, and C. Lei-Fong. 1996. Stoichiometry of carbon, hydrogen, nitrogen, sulfur and oxygen in the particulate matter of the western North Pacific marginal seas. Marine Chemistry 54(2):179-190.

Collins, M.J., G. Muyzer, G.B. Curry, P. Sandberg, and P. Westbroek. 1991. Macromolecules in brachiopod shells: Characterization and diagenesis. Lethaia 24:387-397.

Conte, M.H., J. Volkman, and G. Eglinton. 1994. Lipid biomarkers of the haptophyta. Pp. 351-377, in The Haptophyte Algae, J.C. Green and B.S.C. Leadbeater, eds., Oxford: Clarendon Press.

Coplen, T.B., J.A. Hopple, J.K. Böhlke, H.S. Peiser, S.E. Rieder, H.R. Krouse, K.J.R. Rosman, T. Ding, R.D. Vocke, Jr., K.M. Revesz, A. Lamberty, P. Taylor, and P. DeBievre. 2001. Compilation of minimum and maximum isotope ratios of selected elements in naturally occurring materials and reagents. Water-Resources Investigations Report 01-4222, U.S. Geological Survey, 131 pp.

Coplen, T.B., C. Kendall, and J. Hoppell. 1983. Comparison of stable isotope reference samples. Nature 302:236-238.

Cowie, G.L., and J.I. Hedges. 1996. Digestion and alteration of the biochemical constituents of a diatom (Thalassiosira weissflogii) ingested by an herbivorous zooplankton (Calanus pacificus). Limnology and Oceanography 41:581-594.

Cowie, G.L., and J.I. Hedges. 1992. Improved amino acid quantification in environmental samples: charge-matched recovery standards and reduced analysis time. Marine Chemistry 37:223-238.

Cowie, G.L., and J.I. Hedges. 1984. Carbohydrate sources in a coastal marine-environment. Geochimica et Cosmochimica Acta 48:2075-2087.

Cowie, G.L., J.I. Hedges, and S.E. Calvert. 1992. Sources and relative reactivities of amino acids, neutral sugars, and lignin in an intermittently anoxic marine environment. Geochimica et Cosmochimica Acta 56:1963-1978.

Dauwe, B., and J.J. Middleburg. 1998. Amino acids and hexosamines as indicators of organic matter degradation state in North Sea sediments. Limnology and Oceanography 43:782-798.

deBaar, H.J.W., M. A. Van Leeuwe, R. Scharek, L. Goeyens, K. M. J. Bakker, P. Fritsche. 1997. Nutrient anomalies in Fragilariopsis kergeulensis blooms, iron deficiency and the nitrate/phosphate ratio (A.C. Redfield) of the Antarctic Ocean. Deep-Sea Research II 44(1-2):229-260.

deBaar, H.J.W., J.W. Farrington, and S.G. Wakeham. 1983. Vertical flux of fatty-acids in the North-Atlantic ocean. Journal of Marine Research 41(1):19-41.

Degens, E.T., and K. Mopper. 1976. Factors controlling the distribution and early diagenesis of organic material in marine sediments. Pp. 59-113 in Chemical Oceanography, J.P. Riley and R. Chester, eds., Orlando, Florida: Academic Press.

Deines, P. 1980. The isotopic composition of reduced organic carbon. Pp. 329-406, in Handbook of Environmental Isotope Geochemistry, Vol. 1, F.P. Fontes and J.C. Fontes, eds., University Park, Pennsylvania: Pennsylvania State University Press.

Demaison, G.J., and G.T. Moore. 1980. Anoxic environments and oil source bed genesis. American Association of Petroleum and Geology Bulletin 64(8):1179-1209.

Druffel, E. R.M. 1995. Geochemistry of corals: Proxies of past ocean chemistry, ocean circulation and climate. National Academy of Sciences Colloquium on Carbon Dioxide and Climate Change.

Eglinton, T.I., B.C. Benitez-Nelson, A. Pearson, A.P. McNichol, J.E. Bauer, and E.R.M. Druffel. 1997. Variability in radiocarbon ages of individual organic compounds from marine sediments. Science 277:796-799.

Endo, K., D. Walton, R.A. Reyment, and G.B. Curry. 1995. Fossil intra-crystalline biomolecules of brachiopod shells: Diagenesis and preserved geo-biological information. Organic Geochemistry 13:661-673.

Forch, C., M. Knudson, and S.P. Sorensen. 1902. Reports on the determination of the constants for compilation of hydrographic tables. Det Kongelige Danske videnskabernes selskabs skrifter. Naturvidenskabelig og mathematisk afdeling. 6.Raekke Vol 12(1):1-151.

Freeman, K.H., and J.M. Hayes. 1992. Fractionation of carbon isotopes by phytoplankton and estimates of ancient CO_2 levels. Global Biogeochemical Cycles 6:185-198.

Froelich, P.N. 1980 Analysis of organic carbon in marine-sediments. Limnology and Oceanography 25(3):564-572.

Gagosian, R.B., J.K. Volkman, and G.E. Nigrelli. 1983. The use of sediment traps to determine sterol sources in coastal sediments off Peru. Pp. 369-379 in Advances in Organic Geochemistry 1981, M. Bjoroy et al., eds., New York: John Wiley and Sons.

Gannes, L.Z., D.M. O'Brien, C.M. del Rio. 1997. Stable isotopes in animal ecology: Assumptions, caveats, and a call for more laboratory experiments. Ecology 78(4):1271-1276.

Gélinas, Y., J.A. Baldock, and J.I. Hedges. 2001a. Organic carbon composition of marine sediments: Effect of oxygen exposure on oil generation potential. Science 294 (5540):145-148.

Gélinas, Y., J.A. Baldock, and J.I. Hedges. 2001b. Demineralization of marine and freshwater sediments for CP/MAS ^{13}C NMR analysis. Organic Geochemistry 32(5):677-693.

Gershey, R.M., M.D. MacKinnon, P.J. leB Williams, and R.M. Moore. 1979. Comparison of three oxidation methods used for the analysis of the dissolved organic carbon in seawater. Marine Chemistry 7:289-306.

Goericke, R., J.P Montoya, and B. Fry. 1994. Physiology of isotopic fractionation in algae and cyanobacteria. Pp.187-221 in Stable Isotopes in Ecology and Environmental Science, K. Lajtha and R. H. Michener, eds., Oxford: Blackwell Scientific Publications.

Goericke, R., and B. Fry. 1994. Variations of marine plankton $\delta^{13}C$ with latitude, temperature, and dissolved CO_2 in the world ocean. Global Biogeochemical Cycles 8(1):85-90.

Gonfiantini, R., W. Stichler, and K. Rozanski. 1995. Standards and intercomparison materials distributed by the International Atomic Energy Agency for stable isotope measurements. In: Reference and Intercomparison Materials for Stable Isotopes of Light Elements. International Atomic Energy Agency, Vienna, IAEA-TECDOC-825, pp. 13-29.

Gordon, A.S., and F.J. Millero. 1985. Adsorption mediated decrease in the biodegradation rate of organic compounds. Microbial Ecology 11:289-298.

Gordon, L.I., J.C. Jennings, C.W. Mordy, and A.A. Ross. 1999. To do or not to do? Calibrating WOCE nutrient data. Eos Trans-American Geophysical Union 80 (supplement) OS45.

Gruber, N. 1998. Anthropogenic CO_2 in the Atlantic Ocean. Global Biogeochemical Cycles 12(1):165-191.

Gruber, N., J. L. Sarmiento, and T. F. Stocker. 1996. An improved method for detecting anthropogenic CO_2 in the oceans. Global Biogeochemical Cycles 10(4):809-837.

Gruber, N., and C.D. Keeling. 1999. Seasonal carbon cycling in the Sargasso Sea near Bermuda. In: Bulletin of the Scripps Institution of Oceanography, vol. 30, C. Cox, G.L. Kooyman, and R.H. Rosenblatt, eds. Berkeley, CA: University of California Press, 96 pp.

Guilderson, T.P., D.P. Schrag, M. Kashgarian, and J. Southori. 1998. Radiocarbon variability in the western equatorial Pacific inferred from a high-resolution coral record from Nauru Island. Journal of Geophysical Research, C, Oceans 103(11):24,641-24,650.

Guilderson, T.P., K. Caldeira, and P.B. Duffy. 2000. Radiocarbon as a diagnostic tracer in ocean and carbon cycle modeling. Global Biogeochemical Cycles 14(3):887-902.

Hansell, D.A., and C.A. Carlson, eds. 2002. Biogeochemistry of Marine Dissolved Organic Matter. San Diego, California: Academic Press, 500 pp.

Hansell, D.A. 2001. Determining dissolved organic carbon in the ocean. Oceanography 14:43.

Harvey, H.R., G. Eglinton, S.C.M. O'Hara, and E.D.S. Corner. 1987. Biotransformation and assimilation of dietary lipids by *Calanus* feeding on a dinoflagellate. Geochimica et Cosmochimica Acta 51:3031-3040.

Hayes, J.M. 2001. Fractionation of the isotopes of carbon and hydrogen in biosynthetic processes. Pp. 225-278, in Stable Isotope Geochemistry: Reviews in Mineralogy and Geochemistry, J.W. Valley and D.R. Cole, eds., Washington, D.C.: Mineralogical Society of America.

Hayes, J.M., K.H. Freeman, B.N. Popp, and C.H. Hoham. 1990. Compound-specific isotopic analyses, a novel tool for reconstruction of ancient biogeochemical processes. Organic Geochemistry 16:1115-1128.

Hecky, R.E., K. Mopper, P. Kilham, and E.T. Degens. 1973. The amino acid and sugar composition of diatom cell-walls. Marine Biology 19:323-331.

Hedges, J.I., and R.G. Keil. 1995. Sedimentary organic matter preservation: an assessment and speculative synthesis, ("Invited Discussion Paper"). Marine Chemistry 49:81-115.

Hedges, J.I., and J.H. Stern. 1984. Carbon and nitrogen determinations of carbonate-containing solids. Limnology and Oceanography 29(3):657-663.

Hedges, J.I., J.A. Baldock, Y. Gélinas, C. Lee, M.L. Peterson, and S.G. Wakeham. 2002. The biochemical and elemental compositions of marine plankton: A ^{13}C NMR perspective. Marine Chemistry 78:47-63.

Hedges, J.I., B.A. Bergamaschi, and R. Benner. 1993. Comparative analyses of DOC and DON in natural waters. Marine Chemistry 41:121-134.

Hedges, J.I., J.A. Baldock, Y. Gélinas, C. Lee, M.L. Peterson, and S.G. Wakeham. 2001. Evidence for non-selective preservation of organic matter in sinking marine particles. Nature 409:801-804.

Hedges, J.I., F.S. Hu, A.H. Devol, H.E. Hartnett, and R.G. Keil. 1999. Sedimentary organic matter preservation: A test for selective oxic degradation. American Journal of Science 299:529-555.

Henrichs, S.M., and J.W. Farrington. 1984. Peru upwelling region sediments near 15°S. I. Remineralization and accumulation of organic matter. Limnology and Oceanography 29:1-19.

Henrichs, S.M., J.W. Farrington, and C. Lee. 1984. Peru upwelling region sediments near 15°S – II. Dissolved free and total hydrolyzable amino acids. Limnology and Oceanography 29:20-34.

Hernes, P.J., J.I. Hedges, M.L. Peterson, S.G. Wakeham, and C. Lee. 1996. Neutral carbohydrate geochemistry of particulate material in the central equatorial Pacific. Deep-Sea Research II 43:1181-1204.

Hooker, S. B., H. Clasutre, J. Ras, L. Van Heukelem, C. Targa, R. Barlow, and H. Sessions. 2000. The first SeaWiFS HPLC analysis round-robin experiment (SeaHARRE-1), SeaWiFS Postlaunch Technical Report Series. S.B. Hooker and E. R. Firestone, eds. NASA Technical Memorandum 2000-206892, Vol. 14, 20 pp.

Hut, G. 1987. Consultant's group meeting on stable isotope reference samples for geochemical and hydrological investigations. International Atomic Energy Agency, 42 pp.

Ingalls, A.E., C. Lee, S.G. Wakeham, and J.I. Hedges. (In press). The role of biominerals in the sinking flux and preservation of amino acids in the Southern Ocean along 170°W. Deep-Sea Research II.

Ittekkot, V., E.T. Degens, and S. Honjo. 1984a. Seasonality in the fluxes of sugars, amino acids, and amino sugars to the deep ocean: Panama Basin. Deep Sea Research 31:1071-1083.

Ittekkot, V., E.T. Degens, and S. Honjo. 1984b. Seasonality in the fluxes of sugars, amino acids, and amino sugars to the deep ocean: Sargasso Sea. Deep-Sea Research 31(9):1057-1069.

Jasper, J.P., J.M. Hayes, A.C. Mix, and F.G. Prahl. 1994. Photosynthetic fractionation of ^{13}C and concentrations of CO_2 in the central equatorial Pacific during the last 225,000 years. Paleoceanography 9:781-898.

Jeffrey, S.W., R.F.C. Mantoura, and S.W. Wright, eds. 1997. Phytoplankton Pigments in Oceanography: Monographs on Oceanographic Methodology, UNESCO, 661 pp.

Jenks, P.J., and M. Stoeppler. 2001. The deplorable state of the description of the use of certified reference materials in the literature. Fresenius' Journal of Analytical Chemistry 370(2/3):164-169.

Johnson, K.M., A.G. Dickson, G. Eisheid, C. Goyet, P. Guenther, F.J. Millero, D. Purkerson, C.L. Sabine, R.S. Schottle, D.R.W. Wallace, R.J. Wilke, and C.D. Winn. 1998. Coulometric total carbon dioxide analysis for marine studies: Assessment of the quality of total inorganic carbon measurements made during the U.S. Indian Ocean CO_2 survey 1994-1996. Marine Chemistry 63(1-2):21-37.

Kaiser, K., and R. Benner. 2000. Determination of amino sugars in environmental samples with high salt content by high performance anion exchange chromatography and pulsed amperometric detection. Analytical Chemistry 72(11):2566-2572.

Karl, D.M. 1999. A Sea of Change: Biogeochemical Variability in the North Pacific Subtropical Gyre. Ecosystems 2:181-214.

Karl, D.M., J.R. Christian, J.E. Dore, D.V. Hebel, R.M. Letelier, L.M. Tupas, and C.D. Winn. 1996. Seasonal and interannual variability in primary production and particle flux at Station ALOHA. Deep-Sea Research II 43:1270-1286.

Karlen, I. I.U. Olsson, P. Kallberg, and S. Kilicci. 1964. Absolute determination of the activity of two ^{14}C dating standards. Arkiv for Geofysik Band 4(22):

Keir, R.S., and R.L. Michel. 1993. Interface dissolution control of the ^{14}C profile in marine sediment. Geochimica et Cosmochimica Acta 57:3563-3573.

Kenig, F., J.S. Sinninghe, J.D. de Leeuw, and J.M. Hayes. 1994. Molecular palaeontological evidence for food-web relationships. Naturwissenschaften 81:128-130.

Key, R.M., and S. Rubin. 2002. Separating natural and bomb-produced radiocarbon in the ocean: The potential alkalinity method. Global Biogeochemical Cycles.

Key, R.M., P.D. Quay, P. Schlosser, A.P. McNichol, K.F. von Reden, R.J. Schneider, K.L. Elder, M. Stuiver, and H. Gote Ostlund. In press. WOCE Radiocarbon IV: Pacific Ocean results; P_{10}, $P^{13}N$, $P^{14}C$, P_{18}, P_{19}, and S_4P. Radiocarbon 44(1):239-392.

King, K.J. 1974. Preserved amino acids from slicified protein in fossil Radiolaria. Nature 252:690-692.

Kroopnik, P. 1985. The distribution of ^{13}C of SCO_2 in the world oceans. Deep-Sea Research I 32:57-84.

Kolber, Z.S., C.L. Van Dover, R.A. Niederman, and P. G. Falkowski. 2000. Bacterial photosynthesis in surface waters of the open ocean. Nature 407:177-179.

Ku, T.-L., S. Luo, M. Kusakabe, and J.K.B. Bishop. 1995. [228]Ra-derived nutrient budgets in the upper equatorial Pacific and the role of "new" silicate in limiting productivity. Deep-Sea Research, Pt. II 42(2-3):479-497.

Lamb, M.F., D.L. Sabine, R.A. Feely, R. Wanninkhof, R.M. Key, G.C. Johnson, F.J. Millero, K. Lee, T.-H. Peng, A. Kozyr, J.L. Bullister, D. Greeley, R.H. Byrne, D.W. Chipman, A.G. Dickson, B. Tilbrook, T. Takahashi, D.W.R. Wallace, Y. Watanabe, S. Winn, and C.S. Wong. 2002. Internal consistency and synthesis of Pacific Ocean CO_2 data. Deep-Sea Research, II 49:21-58.

Latasa, M., R.R. Bidigare, M.E. Ondrusek, and M.C. Kennicutt II. 1996. HPLC analysis of algal pigments: A comparison exercise among laboratories and recommendations for improved analytical performance. Marine Chemistry 51:315-324.

Laws, E.A., D.M. Karl, D.G. Redalje, R. S. Jurick, and C.D. Winn. 1983. Variability in ratios of phytoplankton carbon and RNA to ATP and chlorophyll a in batch and continuous cultures. Journal of Phycology 19:439-445.

Laws, E.A., B.N. Popp, R.R. Bidigare, U. Riesbesell, S. Burkhardt, and S.G. Wakeham. 2001. Controls on the molecular distribution and carbon isotopic composition of alkenones in certain haptophyte algae. Geochemistry Geophysics Geosystems January 25, 2001.

Lee, C., and C. Cronin. 1984. Particulate amino acids in the sea: Effects of primary productivity and biological decomposition. Journal of Marine Research 42:1075-1097.

Lee, C., and C. Cronin. 1982. The vertical flux of particulate organic nitrogen in the sea: decomposition of amino acids in the Peru upwelling area of the equatorial Atlantic. Journal of Marine Research 40:227-251.

Lee, C., and S.G. Wakeham. 1988. Organic matter in sea water: Biogeochemical processes. Pp. 1-51 in Chemical Oceanography Vol. 9, J.P. Riley, ed., New York: Academic Press.

Lee, C., S.G. Wakeham, and J.W. Farrington. 1983. Variations in the composition of particulate organic matter in a time-series sediment trap. Marine Chemistry 13:181-194.

Lerperger, M., A.P. McNichol, J. Peden, A.R. Gagnon, K.L. Elder, W. Kutschera, W. Rom, P. Steier. 2000. Nuclear Instruments and Methods in Physics Research 172:501-512.

Levin, I., and V. Hesshaimer. 2000. Radiocarbon—A unique tracer of global carbon cycle dynamics. Radiocarbon 42:69-80.

Lindroth, P., and K. Mopper. 1979. High performance liquid chromatographic determination of subpicomole amounts of amino acids by precolumn fluorescence derivatization with o-phthaldialdehyde. Analytical Chemistry 51:1667-1674.

Lowenstam, H.A., and S. Weiner. 1989. On Biomineralization. New York: Oxford University Press, 324 pp.

Lynch-Stieglitz, J., T.F. Stocker, W.S. Broecker, and R.G. Fairbanks. 1995. The influence of air-sea exchange on (Delta)C-13 of (Sigma)CO_2 in the surface ocean: Observations and modeling. Global Biogeochemical Cycles 9(4):653-665.

Mackensen, A., H.W. Hubberten, N. Scheele, and R. Schlitzer. 1996. Decoupling of delta ^{13}C-CO_2 and phosphate in Recent Weddell Sea deep and bottom water; implications for glacial Southern Ocean paleoceanography. Paleoceanography 11(2):203-215.

MacKinnon, M.D. 1978. A dry oxidation method for the analysis of the TOC in seawater. Marine Chemistry 7:17-37.

MacKinnon, M.D. 1981. The measurement of organic carbon in seawater. Pp.415-443 in Marine Organic Chemistry, E.K. Duursma and R. Dawson, eds., New York: Elsevier.

Marcet, A. 1819. On the specific gravity, and temperature, in different parts of the ocean, and in particular seas; with some account of their saline contents. Philosphical Transactions Royal Society of London 109:161-208.

Madigan, M.T., J.M. Martinko, and J. Parker. 2000. Brock Biology of Microorganisms, 9th ed. Prentice Hall.

Martin, J.H., and R.M. Gordon. 1988. Northeast Pacific iron distributions in relation to phytoplankton productivity. Deep-Sea Research 35:177-196.

Martin, J.H., R.M. Gordon, S.E. Fitzwater, and W.W. Broenkow. 1989. VERTEX: phytoplankton/iron studies in the Gulf of Alaska. Deep-Sea Research 36(5):649-680.

Martin, W.R., A.P. McNichol, and D.C. McCorkle. 2000. The radiocarbon age of calcite dissolving at the seafloor: Estimates from pore water data. Geochimica et Cosmochimica Acta 64(8):1391-1404.

Maslin, M.A., M. A. Hall, N.J. Shackleton, and E. Thomas. 1996. Calculating surface water pCO_2 from foraminiferal organic $\delta^{13}C$. Geochimica et Cosmochimica Acta 60:5089-5100.

Mayer, H., U. Ramadas Bhat, H. Masoud, J. Radziejewska-Lebrecht, C. Widemann, and J.H. Krauss. 1989. Bacterial lipopolysaccharides. Pure Applied Chemistry 67:1271-1282.

Mayer, L.M. 1994. Surface area control of organic carbon accumulation in continental shelf sediments. Geochimica et Cosmochimica Acta 58:1271-1284.

McNichol, A.P., R.J. Schneider, K.F. von Reden, A.R. Gagnon, K.L. Elder, R.M. Key, and P.D. Quay. 2000. Ten years after—the WOCE AMS Radiocarbon Program. Nuclear Instruments and Methods in Physics Research B172:479-484.

Michaels, A.F., D.M. Karl, and D.G. Capone. 2001. Element Stoichiometry: New production and nitrogen fixation. Oceanography 14(4):68-77.

Millero, F.J. 1996. Chemical Oceanography. Boca Raton, Florida: CRC Press, 469 pp.

Millero, F.J., A.G. Dickson, G. Eischeid, C. Goyet, P. Guenther, K.M. Johnson, K. Lee, D. Purkerson, D.L. Sabine, R.G. Schottle, D.R.W. Wallace, R.J. Wilke, and C.D. Winn. 1998. Total alkalinity measurements in the Indian Ocean during the WOCE hydrographic program CO_2 Cruises 1994-1996. Marine Chemistry 63(1-2):9-20.

Moldowan, M.J., J. Dahl, S.R. Jacobson, B.J. Huizinga, F.J. Fago, R. Shetty, D.S. Watt, and K.E. Peters. 1996. Chemostratigraphic reconstruction of biofacies: Molecular evidence linking cyst-forming dinoflagellates with pre-Triassic ancestors. Geology 24:159-162.

Moore, W.S. 2000. Determining coastal mixing rates using radium isotopes. Continental Shelf Research 20(15):1993-2007.

Moore, W.S., J.L. Sarmiento, and R.M. Key. 1986. Tracing the Amazon Component of Surface Atlantic Water using ^{228}Ra, Salinity and Silica. Journal of Geophysical Research 91:2574-2580.

Mopper, K., and K. Larsson. 1978. Uronic and other organic acids in Baltic Sea and Black Sea sediments. Geochimica et Cosmochimica Acta 42:153-163.

Mordy, C.W., M. Aoyama, L.I. Gordon, G.C. Johnson, R.M. Key, A.A. Ross, J.C. Jennings, and J. Wilson. 1999. Deep water comparison studies of the Pacific WOCE nutrient data set. Eos Trans-American Geophysical Union. 80 (supplement), OS43.

National Oceanic and Atmospheric Administration (NOAA). 2001. Repeat Hydrography in Support of the U.S. CLIVAR and CO_2 Programs. Available: www.aoml.noaa.gov/ocd/repeathydro/

National Research Council (NRC). 1993. Applications of Analytical Chemistry to Oceanic Carbon Cycle Studies. Washington, D.C. National Academy Press.

National Research Council (NRC). 2000. Illuminating the Hidden Planet: The Future of Seafloor Observatory Science. Washington, D.C. National Academy Press.

National Research Council (NRC). 1971. Marine Chemistry: A Report of the Marine Chemistry Panel of the Committee on Oceanography. Washington, D.C. National Academy Press.

National Research Council (NRC). 1998. Opportunities in Ocean Sciences: Challenges on the Horizon. Washington, D.C. National Academy Press.

National Science Foundation. 2000. Future of Ocean Chemistry in the U.S. (FOCUS). Arlington, Virginia: National Science Foundation.

National Science Foundation. 2001. Ocean Science at the New Millennium. Arlington, Virginia: National Science Foundation.

Natterer, K. 1892. Chemische untersuchengen im Oestlichen Mittelmeer. Denkschr. Akad. Wiss. Wien 59:53-116.

Nieuwenhuize, J., Y.E.M. Maas, and J.J. Middleburg. 1994. Rapid analysis of organic carbon and nitrogen in particulate materials. Marine Chemistry 45(3):217-224.

Ostlund, H.G. and C.G.H. Rooth. 1990. The North Atlantic Tritium and Radiocarbon Transients 1972-1983. Journal of Geophysical Research—Oceans 95: 20,147-20,165.

Ourisson, G., M. Rohmer, and K. Poralla. 1987. Prokaryotic hopanoids and other polyisoprenoid sterol surrogates. Annual Review of Microbiology 41:310-333.

Parkany, M., and A. Fajgelj. 1999. The use of matrix reference materials in environmental analytical processes. Cambridge, UK: Royal Society of Chemistry.

Parsons, T.R., M. Takahashi, and B. Hargrave. 1977. Biological Oceanographic Processes. 2nd Edition, New York: Pergamon Press, 332 pp.

Paytan, A., W.S. Moore and M. Kastner. 1996. Sedimentation rate as determined by ^{226}Ra activity in marine barite. Geochimica et Cosmochimica Acta 60(22):4313-4319.

Pearson, A., T.I. Eglinton, and A.P. McNichol. 2000. An organic tracer for surface ocean radiocarbon. Paleoceanography 15:541-550.

Pearson, A., A.P. McNichol, G.C. Benitez-Nelson, J.M. Hayes, and T.I. Eglinton. 2001. Origins of lipid biomarkers in Santa Monica Basin surface sediment: A case study using compound-specific ^{14}C analysis. Geochimica et Cosmochimica Acta 65:3123-3137.

Pearson, K. 1901. On lines and planes of closest fit to systems of points in space. Philosophical Magazine Journal of Science 2(6):559-572.

Pederson, T.F., G.B. Shimmield, and N.B. Price. 1992. Lack of enhanced preservation of organic matter in sediments under the oxygen minimum of the Oman Margin. Geochimica et Cosmochimica Acta 56:545-551

Peters, K.E., and J.M. Moldowan. 1993. The Biomarker Guide, Englewood Cliffs, New Jersey: Prentice Hall, 363 pp.

Prahl, F.G., and S.G. Wakeham. 1987. Calibration of unsaturation patterns in long-chain ketone compositions for paleotemperature assessment. Nature 320:367-369.

Prahl, F.G., G. Eglinton, E.D.S. Corner, and S.C.M. O'Hara. 1985. Fecal lipids released by fish feeding on zooplankton. Journal of the Marine Biological Association of the United Kingdom 65:547-560.

Presidents Panel for Ocean Exploration. 2000. (Online): http://oceanpanel.nos.noaa.gov/panelreport/ocean_panel_report.html.

Putter, A.F.R. 1909. Die Ernahrung der Wassertiere und der Stoff haushalt der Gewasser. Jena, Fischer.

Quay, P.D., R. Sonnerup, T. Westby, J. Stutsman, and A.P. McNichol. In press. Anthropogenic changes of the ^{13}C/^{12}C of dissolved inorganic carbon in the ocean as a tracer of CO_2 uptake. Global Biogeochemical Cycles.

Quay, P.D., B. Tilbrook, and C.S. Wong. 1992. Oceanic uptake of fossil fuel CO_2: ^{13}C evidence. Science 256:74-79.

Ratledge, C., and S.G. Wilkinson, eds. 1988. Microbial Lipids. New York: Academic Press.

Raymond, P.A., and J.E. Bauer. 2001. Use of ^{14}C and ^{13}C natural abundances for evaluating riverine, estuarine, and coastal DOC and POC sources and cycling: A review and synthesis. Organic Geochemistry 32(4):469-485.

Redfield, A.C. 1934. On the proportions of derivatives in sea water and their relation to the composition of plankton. Pp. 176-192 in James Johnstone Memorial Volume I., Liverpool: University of Liverpool.

Redfield, A.C., B.H. Ketchum, and F.A. Richards. 1963. The influence of organisms on the composition of seawater. Pp. 26-77 in The Sea. Vol. 2, M.N. Hill, ed.,New York: Interscience.

Robbins, L.L., and K. Brew. 1990. Proteins from the organic matrix of core-top and fossil planktonic foraminifera. Geochimica et Cosmochimica Acta 41:803-810.

Roper, B., S. Burke, R. Lawn, V. Barwick, and R. Walter. 2001. Applications of Reference Materials in Analytical Chemistry. Cambridge: Royal Society of Chemistry.

Ross, A.A., L.I. Gordon, C.W. Mordy, J.C. Jennings, J. Johnson, and J. Wilson. 1999. Nutrient data differences between crossings of WOCE Hydrographic lines. Eos Trans-American Geophysical Union 80 (supplement), OS4.

Rozanski, K., W. Stichler, R. Gonfiantini, E.M. Scott, R.P. Beukens, B. Kromer, and J. van der Plicht. 1992. The IAEA ^{14}C intercomparison exercise 1990. Radiocarbon 34:506-519.

Sabine, C. L., R. M. Key, K. M. Johnson, K. J. Millero, A. Poisoon, J. L. Sarmiento, D. W. R. Wallace, and C. D. Winn. 1999. Anthropogenic CO_2 inventory of the Indian Ocean. Global Biogeochemical Cycles 13(1):179-198.

Sachs, J. P., and D. J. Repeta. 1999. Oligotrophy and nitrogen fixation during the eastern Mediterranean sapropels events. Science 286:2485-2488.

Sarmiento, J.L., T.M.C. Hughes, R.J. Stouffer, and S. Manabe. 1998. Simulated response of the ocean carbon cycle to anthropogenic warming. Nature 393(6682):245-249.

Sarmiento, J., G. Theile, R.M. Key, and W.S. Moore. 1990. Oxygen and nitrate new production and remineralization in the North Atlantic subtropical gyre. Journal of Geophysical Research 95:18303-18315.

Schaule, B.K., and C.C. Patterson. 1981. Lead concentrations in the northeast Pacific: Evidence for global anthropogenic perterbations. Earth Planet Sci. Lett. 54:97-116.

Seki, M.P., J.J. Polovina, R.E. Brainard, R.R. Bidigare, C.L. Leonard, and D.G. Foley. 2001. Biological enhancement at cyclonic eddies tracked with GOES thermal imagery in Hawaiian waters. Geophysical Research Letters 28:1583-1586.

Sharp, J.H. 1993. The dissolved organic carbon controversy: an update. Oceanography 6:45-50.

Sharp, J.H., R. Benner, L. Bennett, C.A. Carlson, S.E. Fitzwater, E.T. Peltzer, and L.M. Tupas. 1995. Analyses of dissolved organic carbon in seawater: The JGOFS EQPAC methods comparison. Marine Chemistry 48:91-108.

Sharp, J.H., C.A. Carlson, E.T. Peltzer, D.M. Castle-Ward, K.B. Savidge, K.R. Rinker. 2002. Final dissolved organic carbon broad community intercalibration and preliminary use of DOC reference materials. Marine Chemistry 77:239-253.

Siegenthaler, U., and J.L. Sarmiento. 1993. Atmospheric carbon dioxide and the ocean. Nature 365(6442):119-125.

Siezen, R.J., and T.H. Mague. 1978. Amino acids in suspended particulate matter from oceanic and coastal waters of the Pacific. Marine Chemistry 6:215-231.

Sigman, D.M., M.A. Altabet, R. Francois, D.C. McCorkle, and G.-F. Gaillard. 1999. The isotopic composition of diatom-bound nitrogen in Southern Ocean sediments. Paleoceanography 14(2):118-134.

Skoog, A., B. Biddanda, and R. Benner. 1999. Bacterial utilization of dissolved glucose in the upper water column of the Gulf of Mexico. Limnology and Oceanography 44(7):1625-1633.

Sonnerup, R.E., P.D. Quay, A.P. McNichol, J.L. Bullister, T.A. Westby, and H.L. Anderson. 1999. Reconstructing the oceanic 13C Suess effect. Global Biogeochemical Cycles 13(4):857-872.

Stichler, W. 1995. Interlaboratory comparison of new materials for carbon and oxygen isotope ratio measurements. Pp. 67-74 in Reference and Intercomparison Materials for Stable Isotopes of Light Elements, Vienna: International Atomic Energy Agency, IAEA-TECDOC-825.

Stoeppler, M., W.R. Wolf, and P.J. Jenks. 2001. Reference Materials for Chemical Analysis, Weinheim, Germany: Wiley-VCH Verlag GmbH.

Stuiver, M., and H. Polach. 1977. Discussion: Reporting of ^{14}C data. Radiocarbon 19:355-363.

Suess, E., and P.J. Müller. 1980. Productivity, sedimentation rate and sedimentary organic matter in the oceans II—Elementary fractionation. Pp. 17-26 in Biogéochimie de la Matière Organique a L'Interface Eau-Sédiment Marin, Paris: CNRS.

Sugimura, Y., and Y. Suzuki. 1988. A high temperature catalytic oxidation method for the determination of non-volatile dissolved organic carbon in seawater by direct injection of a liquid sample. Marine Chemistry 41:105-131.

Summons, R.E., L.L. Jahnke, J.M. Hope, and G.A. Logan. 1999. 2-methyl hopanoids as biomarkers for cyanobacterial oxygenic photosynthesis. Nature 400:554-557.

Takaichi, S., K. Shimada, and J.-I. Ishidsu. 1990. Carotenoids from the aerobic photosynthetic bacterium Erythrobacter longus: β-carotene and its hydroxyl derivatives. Archives of Microbiology 153:118-122.

Takaichi, S., K. Furihata, J.-I. Ishidsu, and K. Shimada. 1991. Carotenoid sulphates from the aerobic photosynthetic bacterium Erythrobacter longus. Phytochemistry 30:3411-3415.

Tanoue, E., N. Handa, and M. Kato. 1982. Horizontal and vertical distributions of particulate organic matter in the Pacific sector of the Antarctic Ocean. Trans. Tokyo University Fisheries 5:65-83.

Toggweiler, J.R., K. Dixon, and K. Bryan. 1989. Simulations of radiocarbon in a coarse-resolution world ocean model 1. Steady state prebomb distributions. Journal of Geophysical Research 94(C6):8217-8242.

Toggweiler, J.R., K. Dixon, and K. Bryan. 1989b. Simulations of radiocarbon in a course resolution world ocean model II: Distribution of bomb-produced ^{14}C. Journal of Geophysical Research 94:8243-8264.

Trees, C.C., D.K. Clark, R.R. Bidigare, and M.E. Ondrusek. 2000. Chlorophyll a versus accessory pigment concentrations within the euphotic zone: A ubiquitous relationship? Limnology and Oceanography 45:1130-1143.

U.S. Federal Agencies with Ocean-Related Programs. 1998. The Year of the Ocean Discussion Papers. Washington, D.C. Office of the Chief Scientist, National Oceanic and Atmospheric Administration.

van Krevelen, D.W. Coal. Amsterdam: Elsevier Publishing, 1961.

Verado, D.J., P.N. Froelich, and A. McIntyre. 1990. Determination of organic carbon and nitrogen in marine sediments using the Carlo Erba NA-1500 Analyzer. Deep-Sea Research 37:157-165.

Volkman, J.K. 1986. A review of sterol markers for marine and terrigenous organic matter. Organic Geochemistry 9:83-99.

Wakeham, S.G., C. Lee, J.W. Farrington, and R.B. Gagosian. 1984. Biogeochemistry of particulate organic matter in the oceans: Results from sediment trap experiments. Deep-Sea Research 31:509-552.

Wakeham, S.G., C. Lee, J.I. Hedges, P.J. Hernes, and M.L. Peterson. 1997. Molecular indicators of diagenetic status in marine organic matter. Geochimica et Cosmochimica Acta 61:5363-5369.

Wakeham, S.G., and E.A. Canuel. 1988. Organic geochemistry of particulate matter in the eastern tropical North Pacific Ocean: Implications for particle dynamics. Journal of Marine Research 46:183-213.

Wakeham, S.G., and C. Lee. 1989. Organic geochemistry of particulate matter in the ocean: The role of particles in oceanic sedimentary cycles. Organic Geochemistry 14:83-96.

Wakeham, S.G., and C. Lee. 1993. Production, transport, and alteration of particulate organic matter in the marine water column. Pp. 145-169 in Organic Geochemistry: Principles and Applications, M.H. Engle and S.A. Macko, eds., New York: Plenum Press.

Weliky, K., E. Suess, C.A. Ungerer, P.J. Muller, and K. Fischer. 1983. Problems with accurate carbon measurement in marine sediments and particulate matter in seawater: A new approach. Limnology and Oceanography 42:192-197.

Whelan, J.K. 1977. Amino acids in surface sediment core of the Atlantic abyssal plain. Geochimica et Cosmochimica Acta 41:803-810.

Williams, P.M. 1992. Measurement of dissolved organic carbon and nitrogen in natural waters. Oceanography 5(2):107-116.

Williams, P.M., and E.R.M. Druffel. 1988. Dissolved organic matter in the ocean: comments on a controversy. Oceanography 1:14-17.

Wright, S.W., S.W. Jeffrey, and R.F.C. Mantoura. 1997. Evaluation of methods and solvents for pigment extraction. Pp. 261-282 in Phytoplankton Pigments in Oceanography, S.W. Jeffrey, R.F.C. Mantoura, and S.W. Wright, eds. Paris: Monographs on Oceanographic Methodology, UNESCO.

Wu, J.F., and E.A. Boyle. 1998. Determination of iron in seawater by high-resolution isotope dilution indutively coupled plasma mass spectrometry after Mg(OH)(2) coprecipitation. Analytica Chimica Acta 367(1-3):183-191.

Wu, J., and E.A. Boyle. 1997. Lead in the western North Atlantic Ocean: Completed response to leaded gasoline phaseout. Geochimica et Cosmochimica Acta 61:3279-3283.

Wu, J., W. Sunda, E.A. Boyle, and D.M. Karl. 2000. Phosphate depletion in the Western North Atlantic Ocean. Science 289:759-762.

Yamamuro, M., and H. Kayanne. 1995. Rapid direct determination of organic carbon and nitrogen in carbonate-bearing sediments with a Yanaco MT-5 CHN analyzer. Limnology and Oceanography 40(5):1001-1005.

Zhang, J.-Z., and P.B. Ortner. 1998. Effect of thawing conditions on the recovery of reactive silicic acid from frozen natural water samples. Water Research 32:2553-2555.

Zhang, J.-Z., C. Fischer, and P.B. Ortner. 1999. Dissolution of silicate from glassware as a contaminant in silicate analysis of natural water samples. Water Research 33:2879-2883.

Zhang, J.-Z., C.W. Mordy, L.I. Gordon, A. Ross, H. Garcia, M. Pahlow, and U. Riebesell. 2000. Temporal trends in deep ocean redfield ratios. Science 289:1839a.

Zschunke, A., ed. 2000. Reference Materials in Analytical Chemistry: A Guide for Selection and Use, Vol. 40. Berlin: Springer-Verlag.

APPENDIX
A

Committee and Staff Biographies

COMMITTEE MEMBERS

Andrew Dickson *(Chair)* is an Associate Professor-in-Residence at the Scripps Institution of Oceanography. His research focuses on the analytical chemistry of carbon dioxide in sea water, biogeochemical cycles in the upper ocean, marine inorganic chemistry, and the thermodynamics of electrolyte solutions at high temperatures and pressures. His expertise lies in the quality control of oceanic carbon dioxide measurements and in the development of underway instrumentation for the study of upper ocean biogeochemistry. Dr. Dickson served on the NRC Committee on Oceanic Carbon. He is presently a member of the IOC CO_2 Advisory Panel and of the PICES Working Group 13 on CO_2 in the North Pacific.

Robert Bidigare is a Professor in the Department of Oceanography at the University of Hawaii. His research focuses on bio-optical oceanography, nutrient cycling, phytoplankton pigment biochemistry, and the intermediary metabolism of marine plankton. Of particular importance to this committee is Dr. Bidigare's research exploring the ability to identify and quantify the pigments in phytoplankton, which is essential for estimating ocean primary production from satellite observations. Dr. Bidigare has served on the Joint Task Group for Standard Methods for the Examination of Water and Wastewater, IOC Group of Experts on Standards and Reference Materials, and was the U.S. algal pigment consultant for the international JGOFS program. In addition, Dr. Bidigare coordinated an international pigment inter-comparison exercise among JGOFS pigment analysts.

John Hedges was a Professor of Chemical Oceanography at the University of Washington. His research focused on marine organic geochemistry, specifically the sources, transport, and fates of organic materials in aquatic environments and the comparative geochemistry of carbohydrates, lignins, and proteins. His expertise was in the analysis of organic compounds in seawater. Dr. Hedges participated in the Carbon in the Amazon River Experiment (CAMREX) and the Joint Global Ocean Flux Study (JGOFS). Furthermore, Dr. Hedges served on the NRC Committee on Oceanic Carbon and was a member of the Ocean Studies Board. He passed away before this report was released.

Kenneth Johnson is a Senior Scientist at the Monterey Bay Aquarium Research Institute. His research interests are focused on the development of new analytical methods for chemicals in seawater and application of these tools to studies of chemical cycling throughout the ocean. His group has developed a variety of analytical methods for analyzing metals present at ultratrace concentrations in seawater. His expertise lies in trace metal analysis and instrumentation. The creation of reference materials to calibrate these instruments is important for the production of long-term, high-precision datasets. Dr. Johnson has participated on the NRC Committee on Marine Environmental Monitoring and the Marine Chemistry Study Panel.

Denise LeBlanc is the Group Leader for the Marine Sciences Group and the Manager of the Certified Reference Materials Program at the Institute for Marine Biosciences (IMB) of the National Research Council of Canada. The Certified Reference Materials Program manufactures instrument calibration standards and certified reference materials for shellfish toxins, PCBs, PACs, and trace elements in marine sediments, in biological tissues, and in seawater. Her experience resides in the manufacture and long-term production of reference materials.

Cindy Lee is a Professor at the Marine Sciences Research Center of Stony Brook University. Dr. Lee's research examines the distribution and behavior of biogenic organic compounds, in particular the rates and mechanisms of transformation reactions occurring as these compounds undergo alteration. Her research investigates organic compounds in the sediments and waters of open ocean and coastal areas, salt marshes, lakes, as well as the atmosphere above these areas. Her expertise centers on the analytical techniques used to measure organic matter in the ocean. Dr. Lee is currently a member of the Ocean Studies Board and was a member of the U.S. Scientific Committee on Problems of the Environment (SCOPE).

Frank Millero is a Professor of Marine and Physical Chemistry and Associate Dean at the Rosenstiel School of Marine and Atmospheric Science at the University of Miami. Dr. Millero's research interests include the application of physical chemical principles to natural waters to understand how ionic interactions affect the thermodynamics and kinetics of processes occurring in the oceans. He has extensive experience with many aspects of marine chemistry and chemical analysis including the analysis of trace metals and gases in seawater. Dr. Millero is a former member of the NRC's Ocean Studies Board and was a member of the Study Committee on Effects of Human Activities on the Coastal Ocean.

James Moffett is an Associate Scientist of Marine Chemistry and Geochemistry at the Woods Hole Oceanographic Institution. His research interests include the speciation and redox chemistry of trace elements in natural waters, the analysis and characterization of biologically produced chelators, and the interactions between metal geochemistry and cyanobacterial ecology. He is particularly interested in the relationship between speciation and analysis for analytes exhibiting dynamic chemistry in the seawater matrix (e.g., non-stable redox states). Dr. Moffett has served on numerous national and international committees and chaired a recent meeting to establish protocols for iron standards in seawater formed under the auspices of SCOR Working Group 109.

Willard Moore is a Professor of Geochemistry and Chemical Oceanography at the University of South Carolina. Dr. Moore's research focuses on the use of natural radioisotopes as tracers of geologic and oceanographic processes, such as the interactions of river water and sediments with sea water; flow of ground water through salt marshes; the mixing rate of the ocean; hydrothermal processes at ocean spreading centers; the internal structure of minerals; the ages, rates, and processes of formation of manganese nodules; the rate of growth of corals; and sea level changes. Dr. Moore has served on the NSF Future of Ocean Chemistry Steering Committee and the SCOR Groundwater Discharge Working Group.

Ann McNichol is a Research Specialist at the National Ocean Sciences Accelerator Mass Spectrometry Facility at the Woods Hole Oceanographic Institution, which produces high-precision ^{14}C measurements from small-volume samples. Dr. McNichol's research interests include the study and use of carbon, nitrogen, and oxygen isotope techniques to quantify biogeochemical processes, the study of the fate of organic matter (both natural and anthropogenic) in sediments, and the development of techniques for analysis of oceanographic samples by AMS.

Edward Peltzer is a Senior Research Specialist at the Monterey Bay Aquarium Research Institute. Dr. Peltzer's research focuses on the development of new analytical techniques for the measurement of major components and selected trace gases in natural and synthetic clathrate hydrates. His past research investigated the role of dissolved organic matter in the global ocean carbon cycle emphasizing the development of new techniques for the measurement of dissolved organic carbon and nitrogen. He collaborated with several other U.S. JGOFS investigators on the development of ad hoc standard reference materials and blanks for these substances.

Stan Van Den Berg is a Professor of Chemical Oceanography at the University of Liverpool. His research interests focus on the chemical speciation of trace elements and organic compounds in natural waters and the redox chemistry of metals and sulfides. His research group has pioneered advances in analytical techniques using electroanalytical methods (cathodic stripping voltammetry and chronopotentiometry). Dr. Van Den Berg is a broad-based analytical chemist.

NATIONAL RESEARCH COUNCIL STAFF

Joanne Bintz (Study Director) has been a Program Officer at the Ocean Studies Board since June 2001. She received her Ph.D. in Biological Oceanography from the University of Rhode Island Graduate School of Oceanography. Dr. Bintz has conducted research on the effects of decreasing water quality on eelgrass seedlings and the effects of eutrophication on shallow macrophyte dominated coastal ponds using mesocosms. Her interests include coastal ecosystem ecology and function, eutrophication of coastal waters, seagrass ecology and restoration, oceanography and global change education, and coastal management and policy.

Darla Koenig (Senior Project Assistant) received her B.A. in English and her M.Hum. in Humanities from the University of Dallas in 1992 and 1997, respectively. During her tenure with the Ocean Studies Board, she has worked on studies involving living marine resources, fisheries issues, and marine chemistry.

APPENDIX
B

Workshop Participants

Lihini Aluwihare, Scripps Institution of Oceanography, La Jolla, California
Ginger Armbrust, University of Washington, Seattle
Ed Boyle, Massachusetts Institute of Technology, Boston
Lloyd Currie, NIST, Gaithersburg, Maryland
Jack Fassett, NIST, Gaithersburg, Maryland
Dennis Hansell, Rosenstiel School of Marine and Atmospheric Science, University of Miami, Florida
Kai Uwe Hinrichs, Woods Hole Oceanographic Institution, Woods Hole, Massachusetts
Ken Inn, NIST, Gaithersburg, Maryland
Marv Lilley, University of Washington, Seattle
George Luther, University of Delaware, Dover
Peter Milne, National Science Foundation, Arlington, Virginia
Ann Pearson, Harvard, Cambridge, Massachusetts
Kathleen Ruttenberg, Woods Hole Oceanographic Institution, Woods Hole, Massachusetts
Eric Saltzman, University of California, Irvine
Peter Santschi, Texas A&M, College Station
Thomas Torgersen, University of Connecticut, Groton
Chuck Trees, NASA Goddard Space Flight Center, Greenbelt, Maryland
Stuart Wakeham, Skidaway Institute of Oceanography, Savannah, Georgia
Scott Willie, NRC-Canada, Ottawa
Jia-Zhong Zhang, Rosenstiel School of Marine and Atmospheric Science, University of Miami, Florida

APPENDIX
C

Glossary

Accelerator mass spectrometry: Instrumental technique for direct enumeration of low-level radioactive nuclides.

Accuracy: The closeness of agreement between a test result and the accepted reference value.

Algaenans: Hydrolysis resistant macromolecules of algal origin.

Alkenones: Long-chain, unsaturated ketones that are synthesized exclusively by certain haptophyte microalgae (e.g., *Emiliania huxleyi*).

Analyte: The substance being measured in an analytical procedure.

C3 photosynthetic pathway: Also known as the Calvin cycle. A series of enzymatically mediated photosynthetic reactions during which CO_2 is reduced to 3-phosphoglyceraldehyde and the CO_2 acceptor, ribulose 1,5-bisphosphate, is regenerated.

C4 photosynthetic pathway: The set of reactions through which CO_2 is fixed to a compound known as phosphoenolpyruvate (PEP) to yield oxaloacetate, a four-carbon compound.

CAM (Crassulacean Acid Metabolism) photosynthetic pathway: A variant of the C_4 pathway; phosphoenolpyruvate fixes CO_2 in C_4 compounds at night, and then, the fixed CO_2 is transferred to the ribulose bisphosphate of the Calvin cycle within the same cell during the day. Characteristic of most succulent plants, such as cacti.

Carotenoids: Organic molecules that function as accessory pigments in phytoplankton photosynthesis.

Certified Reference Material: Reference material, accompanied by a certificate, one or more of whose property values are certified by a procedure that establishes its traceability to an accurate realization of the unit in which the property values are expressed, and for which each certified value is accompanied by an uncertainty statement at a specified level of confidence.

Certified Value: For a certified reference material, the value that appears in the certificate accompanying the material.

Chromatography: Analytical technique to separate chemical species by continuous partitioning between a mobile and stationary phase.

Colloids: Material in the nanometer to micrometer size range whose characteristics and reactions are largely controlled by surface properties.

Consensus Value (of a given quantity): For a reference material, the value of the quantity obtained by interlaboratory testing, or by agreement between appropriate bodies or experts.

Cosmogenic Nuclides: Isotopes of elements produced by the action of cosmic rays (often radioactive).

Deltaic: Of or pertaining to river deltas.

Diatoms: Fucoxanthin containing phytoplankton characterized by the presence of a silica-impregnated cell wall.

Dinoflagellates: Peridinin-containing phytoplankton characterized by a "rotating" swimming movement produced by the combined action of their flagellae.

Fallout or bomb-produced nuclides: Isotopes of elements produced from nuclear reactors or by detonation of nuclear devices (often radioactive).

Foraminifera: Micrometer-size animals that possess calcium carbonate tests.

Haptophyte: Hex-fuxoxanthin–containing phytoplankton whose representatives include calcium carbonate plate-forming species (e.g., *E. huxleyi*).

Heterotrophic activity: A mode of nutrition based on the oxidation of organic matter.

Homogeneity: Condition of being of uniform structure or composition with respect to one or more specified properties. A reference material is said to be homogeneous with respect to a specified property if the property value, as determined by tests on samples of specified size, is found to lie within the specified uncertainty limits, the samples being taken either from different supply units (bottles, packages, etc.) or from a single supply unit.

Hopanoids: Membrane stabilizing chemical components found in the cell membranes of bacteria.

Hydrolysis-resistant biomacromolecules: Large organic compounds of biological origin that can not be broken into subunits by heating with acid or base.

Hydrous Minerals: Minerals that contain water within their matrices.

Hygroscopic Salts: Salts that have a high affiinity for water.

Information Value: Value noted for the concentration of a substance in a reference material, however insufficient information exists to assess the associated uncertainty. Nevertheless, it is believed that this information is of substantial interest to potential users of the reference material.

Interlaboratory Test: Series of measurements of one or more quantities performed independently by a number of laboratories on samples of a given material.

Isotope: Two or more species of atoms of a chemical element with the same atomic number and different atomic mass or mass number.

Isotope Dilution: Condition that occurs when a stable isotope of the analyzed metal is added to the seawater. After equilibrium with the sample, the isotope can be expected to behave in the same way as the metal being determined, and can be used to establish the recovery of the preconcentration steps used prior to detection. Instead of the absolute quantity, the isotopic ratio is determined.

Labile: Prone to reaction or easily degraded.

Lyophilization: Freeze-drying.

Macromolecular: Pertaining to large organic molecules.

Macronutrients: Elements or compounds (like nitrate, phosphate, and silicate) that are necessary to sustaiin life and that are typically present in larger amounts than the micronutrients.

Matrix: The immediate environment (milieu) surrounding an element or compound. For example, seawater, sediment, and particulate material are general matrices of interest in this report. Also, organic matter, opal, carbonate and aluminosilicate are more specific solid-phase matrices of interest.

Micronutrients: Elements or compounds (like iron, other trace metals, and vitamins) that are necessary to sustain life and that are present in smaller amounts than the macronutrients (see above).

Mycosporine-like Amino Acids (MAA): Small molecular weight organic compounds that absorb UV light and are thought to function as natural sunscreens.

Non-bioactive metals: Elements that do not take an active role in biological cycles.

Paleoenvironmental Indicators: Elements or compounds that reflect ocean history.

Paleovegetation Indicators: Elements or compounds that reflect historic vegetation patterns.

Photoautotrophic: Mode of nutrition based on the use of solar energy to synthesize organic compounds.

Photobioreactors: Light-controlled chambers used to grow microalgae in mass culture.

Photoheterotrophy: A mode of nutrition based on sunlight and the oxidation of organic compounds.

Phytoplankton: Small (often microscopic) aquatic plants suspended in water.

Plant Chars: A form of black carbon obtained by burning plant materials.

Polymerization: The act of forming a larger molecule from a combination of smaller structural units.

Polynuclear Aromatic Hydrocarbons (PAH): Organic molecules composed of 2 or more fused benzene rings.

Precision: The closeness of agreement between independent test results obtained under prescribed conditions.

Primary Standard: Standard that is designated or widely acknowledged as having the highest metrological qualities and whose value is accepted without reference to other standards of the same quantity, within a specified context.

Protists: Single-celled organisms more complex than bacteria that include protozoans and some types of algae.

Radionuclides: Isotopes of elements characterized by spontaneous nuclear interconversions.

Reference Material: Material or substance one or more of whose property values are sufficiently homogeneous and well established to be used for the calibration of an apparatus, the assessment of a measurement method, or for assigning values to materials.

Reference Method: Thoroughly investigated method, clearly and exactly describing the necessary conditions and procedures for the measurement of one or more property values that has been shown to have accuracy and precision commensurate with its intended use and that can therefore be used to assess the accuracy of other methods for the same measurement, particularly in permitting the characterization of a reference material.

Salinity: A measure of the salt content of seawater (approximately the weight (g) of the dissolved inorganic matter in 1 kg of seawater.

Secondary Standard: Standard whose value is assigned by comparison with a primary standard of the same quantity.

Secular Equilibrium: A condition that occurs when a chain of radionuclides has reached a steady state condition, in which the rate of decay of daughter nuclides is balanced by their rate of formation by decay of each parent. In this condition, the radioactivity (measured in disintigrations per minute) of each radionuclide in a chain is the same.

Spectroscopic Analysis: Measurement of chemical species based on the detection of electromagnetic radiation.

Stability: Ability of a reference material, when stored under specified conditions, to maintain a stated property value within specified limits for a specified period of time.

Traceability: Property of the result of a measurement or the value of a standard whereby it can be related, with a stated uncertainty, to stated references, usually national or international standards, through an unbroken chain of comparisons.

Uncertainty of a Certified Value: Estimate attached to a certified value of a quantity which characterizes the range of values within which the "true value" is asserted to lie with a stated level of confidence.

Uncertified Value: Value of a quantity, included in the certificate of a CRM or otherwise supplied, which is provided for information only but is not certified by the producer or the certifying body. (Also see Information Value).

Zooplankton: Assemblage of drifting or feebly swimming animals that are often minute in size.

APPENDIX
D
Acronym List and Chemical Terminology

ACRONYM LIST

AESOPS   Antarctic Environment and Southern Ocean Process Study

AMS   Accelerator Mass Spectrometry

AOU   Apparent Oxygen Utilization

BATS   Bermuda Atlantic Time Series

CHN   Carbon Hydrogen Nitrogen

CSIA   Compound-specific Isotopic Analysis

CSK   Cooperative Study of the Kuroshivo

DIC   Dissolved Inorganic Carbon

DOC   Dissolved Organic Carbon

DOM   Dissolved Organic Matter

DOP   Dissolved Organic Phosphorus

EqPac   Equatorial Pacific

FIRI   Fourth International Radiocarbon Intercomparison

GFAAS   Graphite Furnace Atomic Absorption Spectroscopy

irm-GCMS   Gas chromatography-isotope ratio mass spectrometry

HOT   Hawaii Ocean Time-series
HPLC   High power liquid chromatography

IAEA   International Atomic Energy Agency
ICP-MS   Inductively-Coupled Plasma Mass Spectrometry

JGOFS   Joint Global Ocean Flux Study

NIST   National Institute of Standards and Technology
NMR   Nuclear Magnetic Resonance
NOAA   National Oceanic and Atmospheric Administration
NRC   National Research Council
NSF   National Science Foundation

PAH   Polynuclear Aromatic Hydrocarbons
PCB   Polychlorinated biphenyl
POC   Particulate Organic Carbon
POM   Particulate Organic Matter

SCOR   Scientific Committee on Ocean Research
SeaWiFS   Sea-viewing Wide Field-of-View Sensor

USGS   United States Geological Survey

WOCE   World Ocean Circulation Experiment

NUCLIDES

^{227}Ac	Actinium-227
^{26}Al	Aluminum-26
^{241}Am	Americium-241
^{39}Ar	Argon-39
^{7}Be	Beryllium-7
^{10}Be	Beryllium-10
^{12}C	Carbon-12
^{13}C	Carbon-13
^{14}C	Carbon-14
^{137}Cs	Cesium-137
^{36}Cl	Chlorine-36
^{129}I	Iodine-129
^{85}Kr	Krypton-85
^{14}N	Nitrogen-14
^{15}N	Nitrogen-15
^{210}Pb	Lead-210
^{239}Pu	Plutonium-239
^{240}Pu	Plutonium-240
^{210}Po	Polonium-210
^{231}Pa	Protactinium-231
^{223}Ra	Radium-223
^{224}Ra	Radium-224
^{226}Ra	Radium-226
^{228}Ra	Radium-228
^{222}Rn	Radon-222
^{90}Sr	Strontium-90
^{228}Th	Thorium-228
^{230}Th	Thorium-230
^{232}Th	Thorium-232
^{234}Th	Thorium-234
^{3}H	Tritium
^{234}U	Uranium-234
^{235}U	Uranium-235
^{238}U	Uranium-238

COMPOUNDS

$CaCO_3$	calcium carbonate
CO_2	carbon dioxide
CO	carbon monoxide
DMS	dimethyl sulfide
N_2	gaseous nitrogen
HCl	hydrochloric acid
HF	hydrofluoric acid
$Mg(OH)_2$	magnesium hydroxide
CH_4	methane
NO_3	nitrate
N_2O	nitrous oxide
PO_4	phosphate
H_3PO_4	phosphoric acid
KCl	potassium chloride
$Si(OH)_4$	silicate
$AgBr$	silver bromide
$AgCl$	silver chloride
$AgNO_3$	silver nitrate
$NaOH$	sodium hydroxide
H_2SO_4	sulfuric acid
H_2SO_3	sulfurous acid
H_2O	water

FORMULAS FOR CALCULATING $\delta^{13}C$, $\delta^{14}C$, and $\Delta^{14}C$
(Units in parts per thousand or per mille — ‰)

$$\delta^{13}C = \left[\frac{(^{13}C / ^{12}C)_{sample}}{(^{13}C / ^{12}C)_{standard}} - 1 \right] \times 1000$$

$$\delta^{14}C = \left[\frac{(^{14}C / ^{12}C)_{sample}}{(^{14}C / ^{12}C)_{standard}} - 1 \right] \times 1000$$

$$\Delta^{14}C = \delta^{14}C - \left[2(\delta^{13}C + 25 \right] \times \left[1 + \frac{\delta^{14}C}{1000} \right]$$

Reference Materials Listed within this Report

Any current comprehensive, printed, list of reference materials for ocean science would quickly become obsolete as new materials are produced and others go out of stock. Perhaps the most complete list produced to date is that assembled by Dr. Adrianna Cantillo of the NOAA Status and Trends Program:

Cantillo, A.Y. 1995. Standard and Reference Materials for Environmental Science (Part 1 and Part 2). NOAA Technical Memo. 94. NOAA/NOS/ORCA, Silver Spring, MD. 752 pp.

Another list of environmentally relevant reference materials is maintained at the IAEA web-site (http://www.iaea.or.at/programmes/nahunet/e4/nmrm/). Nevertheless, new materials are produced each year, and older ones run out and are not replaced so it is important to keep such lists current.

Clearly this is an arena in which computer database technology has a major role to play. Perhaps the best developed such database is COMAR (an index Code of Reference Materials). This database tries to provide an up-to-date way to locate CRMs from all around the world. The Central Secretariat is presently housed at the Federal Institute for Materials Research and Testing (BAM), Berlin, Germany. Further information is available at the COMAR website (http://www.comar.bam.de/). Nevertheless, this database is not restricted to ocean science applications (indeed they represent a very small fraction of available reference materials), and is not particularly user-friendly.

There is thus a need for a searchable, ocean science specific, database of currently available reference materials, and one of the recommendations of this report is that such a database be developed and maintained by the ocean science community.

Reference materials cited in this report, currently available for use in ocean science.

Identifier	Matrix	Analyzed For	Source (see footnotes)
Ocean CO_2 RMs	Natural Seawater	Total Dissolved Inorganic carbon, Total Alkalinty	Dr. A. G. Dickson, U.C. San Diego[1]
DOC-CRM	Natural Seawater	Dissolved Organic Carbon	Dr. D. Hansell, U. Miami[2]
FIRI A, B (†)	Kauri Wood	pMC	Dr. E. M. Scott, U. Glasgow [3]
FIRI C	Turbidite Carbonate	pMC	Dr. E. M. Scott, U. Glasgow
FIRI E	Humic Acid	pMC	Dr. E. M. Scott, U. Glasgow
FIRI G, J	Barley Mash	pMC	Dr. E. M. Scott, U. Glasgow
FIRI H	German Oak Dendro-dated Wood	pMC	Dr. E. M. Scott, U. Glasgow
FIRI I	Belfast Scots Pine Dendro-dated Wood Cellulose	pMC	Dr. E. M. Scott, U. Glasgow
FIRI K	Cambridge Dendro-wood Whole Wood & Cellulose	pMC	Dr. E. M. Scott, U. Glasgow
IAEA-315 (*)	Marine Sediment	Radionuclides	IAEA [4]
IAEA-356 (*)	Polluted Marine Sediment	Trace Elements & Methyl Mercury	IAEA
IAEA-368 (*)	Marine Sediment	Radionuclides	IAEA
IAEA-381	Natural Seawater	Radionuclides	IAEA
IAEA-383 (*)	Marine Sediment	Organic Contaminants	IAEA
IAEA-384 (*)	Lagoon Sediment	Radionuclides	IAEA (available soon)
IAEA-408 (*)	Marine Sediment	Organic Contaminants	IAEA

Identifier	Matrix	Analyzed For	Source (see footnotes)
IAEA-C1	Carrara Marble	$\delta^{13}C$, pMC	IAEA
IAEA-C2	Freshwater Travertine	$\delta^{13}C$, pMC	IAEA
IAEA-C3	Cellulose, 1989 Growth of Tree	$\delta^{13}C$, pMC	IAEA
IAEA-C5	Sub-fossil Wood E. Wisconsin Forest	$\delta^{13}C$, pMC	IAEA
IAEA-C6	Sucrose (ANU)	$\delta^{13}C$, pMC	IAEA
IAEA-CO-1	$CaCO_3$ (Carrara Marble)	$\delta^{13}C$	IAEA
IAEA-CO-8 (IAEA-KST)	$CaCO_3$ (Calcite)	$\delta^{13}C$	IAEA
IAEA-CO-9 (IAEA-NZCH) LSVEC	Barium Carbonate	$\delta^{13}C$	IAEA
SRM 8455	Lithium Carbonate	$\delta^{13}C$	IAEA (out of stock at NIST)
SRM 1648 (*)	Urban Atmospheric Particulate	Trace Elements	NIST[5]
SRM 1649A (*)	Urban Dust	Organic Constituents	NIST
SRM 1650A (*)	Diesel Particulate	Organic Constituents	NIST (being replaced)
SRM 1939 (*) NIST	River Sediment	PCBs & Chlorinated Pesticides	
SRM 1941B (*)	Marine Sediment	Various Organics	NIST (in prep.)
SRM 1944 (*)	New York/New Jersey Waterway Sediment	Various Organics	NIST
SRM 1945 (*)	Whale Blubber	PCBs & Chlorinated Pesticides	NIST
SRM 4321C	1 M HNO_3	^{238}U, ^{235}U, ^{234}U	NIST
SRM 4325	$BeCl_2$ solution	$^{10}Be/^9Be$	NIST
SRM 4339B	1.4 M HCl ?	^{228}Ra	NIST (in prep.)
SRM 4350B	River Sediment	Radionuclides	NIST
SRM 4357	Marine Sediment	Radionuclides	NIST
SRM 4361C	Water	^3H	NIST
SRM 4969	1.5 M HCl	^{226}Ra	NIST
SRM 4990C	Oxalic Acid Powder	^{14}C	NIST
RM 8452	Sucrose (ANU)	$\delta^{13}C$, pMC	NIST
SRM 8454 NBS 19 (TS-limestone)	$CaCO_3$	$\delta^{13}C$, $\delta^{18}O$	NIST, IAEA
RM 8539 NBS 22	Oil	δ^2H, $\delta^{13}C$	NIST, IAEA

Identifier	Matrix	Analyzed For	Source (see footnotes)
RM 8540 PEF1 IAEA-CH-7	Polyethylene	δ^2H, $\delta^{13}C$	NIST, IAEA
RM 8541 USGS24	C (graphite)	$\delta^{13}C$	NIST, IAEA
RM 8562	CO_2 gas	$\delta^{13}C$	NIST, IAEA
RM 8563	CO_2 gas	$\delta^{13}C$	NIST, IAEA
RM 8564	CO_2 gas	$\delta^{13}C$	NIST, IAEA
IAEA-NGS1	Natural Gas (Coal)	δ^2H, $d^{13}C$	NIST-ACG[6]
IAEA-NGS2	Natural Gas (Oil)	δ^2H, $d^{13}C$	NIST-ACG
IAEA-NGS3 NIST-ACG	Natural Gas (Biogenic)	δ^2H, $\delta^{13}C$	
HISS-1 (*)	Marine Sediment	Trace Metals	NRC-IMB, Canada[7]
HS-3B (*)	Harbor Sediment	Polycyclic Aromatic Hydrocarbons	NRC-IMB, Canada
HS-4B (*)	Harbor Sediment	Polycyclic Aromatic Hydrocarbons	NRC-IMB, Canada
HS-5 (*)	Harbor Sediment	Polycyclic Aromatic Hydrocarbons	NRC-IMB, Canada
HS-6 (*)	Harbor Sediment	Polycyclic Aromatic Hydrocarbons	NRC-IMB, Canada
SES-1 (*)	Estuarine Sediment	Polycyclic Aromatic Hydrocarbons	NRC-IMB, Canada
CARP-2 (*)	Ground Whole Carp	Organochlorine Compounds	NRC-INMS, Canada[8]
DOLT-2 (*)	Dogfish Liver	Trace Metals	NRC-INMS, Canada
DORM-2 (*)	Dogfish Muscle	Trace Metals	NRC-INMS, Canada
LUTS-1 (*)	Non-defatted Lobster Hepatopancreas	Trace Metals	NRC-INMS, Canada
MESS-3 (*)	Marine Sediment	Trace Metals	NRC-INMS, Canada
NASS-5	Natural Seawater	Trace Metals	NRC-INMS, Canada
PACS-2 (*)	Marine Sediment	Trace Metals	NRC-INMS, Canada
TORT-2 (*)	Lobster Hepatopancreas	Trace Metals	NRC-INMS, Canada
IAPSO Standard Seawater	Natural Seawater	Salinity	Ocean Scientific International Ltd.[9]
CSK Standard Solutions (various)	3.05% NaCl Solution	Nitrate, nitrite, phosphate, silicate	Wako Pure Chemical Industries[10]
CSK Standard Solution	Potassium Iodate (0.0100N)	For oxygen titration	Wako Pure Chemical Industries
MAG-1 (*)	Marine Sediment	Elemental Composition	USGS[11]

Table Footnotes

†4th International Radiocarbon Intercomparison (Bryant et al., 2001; Boaretto et al., in press): http://www.stats.gla.ac.uk/~marian/intercomp.html
*These RMs were suggested in Table 4.2 as materials that could be useful for organic analyses (though not explicitly designed for such constituents).

Sources for RMs mentioned in Table:
(1) Dr. Andrew Dickson: http://www-mpl.ucsd.edu/people/adickson/CO2_QC/ (Contact: co2rms@mpl.ucsd.edu)
(2) Dr. Dennis Hansell: http://www.bbsr.edu/Labs/hanselllab/crm.html (Contact: wenchen@rsmas.miami.edu)
(3) Prof E M Scott, Dept of Statistics, University of Glasgow, Glasgow G12 8QW UK (marian@stats.gla.ac.uk)
(4) IAEA: http://www.iaea.org/programmes/aqcs/
(5) NIST: http://ts.nist.gov/ts/htdocs/230/232/232.htm
(6) NIST Atmospheric Chemistry Group: http://www.cstl.nist.gov/div837/837.01/outputs/standards/StdMat.htm
(7) NRC-IMB, Canada: http://www.imb.nrc.ca/crmp_e.html
(8) NRC-INMS, Canada: http://www.ems.nrc.ca/ems1.htm
(9) Ocean Scientific: http://www.oceanscientific.com/seawaterdivision.htm
(10) Although these were originally distributed by the Sagami Chemical Research Center, they are now available from Wako Pure Chemical Industries through their on-line catalog (Keyword: CSK): http://www.wako-chem.co.jp/specialty/
(11) U.S. Geological Survey: http://minerals.cr.usgs.gov/geo_chem_stand/

Additional sources for Reference Materials:

(1) A limited selection of environmental reference materials is also available from the European Institute for Reference Materials and Measurements in Belgium: http://www.irmm.jrc.be/

(2) International Steering Committee for Black Carbon Reference http://www.du.edu/~dwismith/bcsteer.html